OUT OF THE FIERY FURNACE

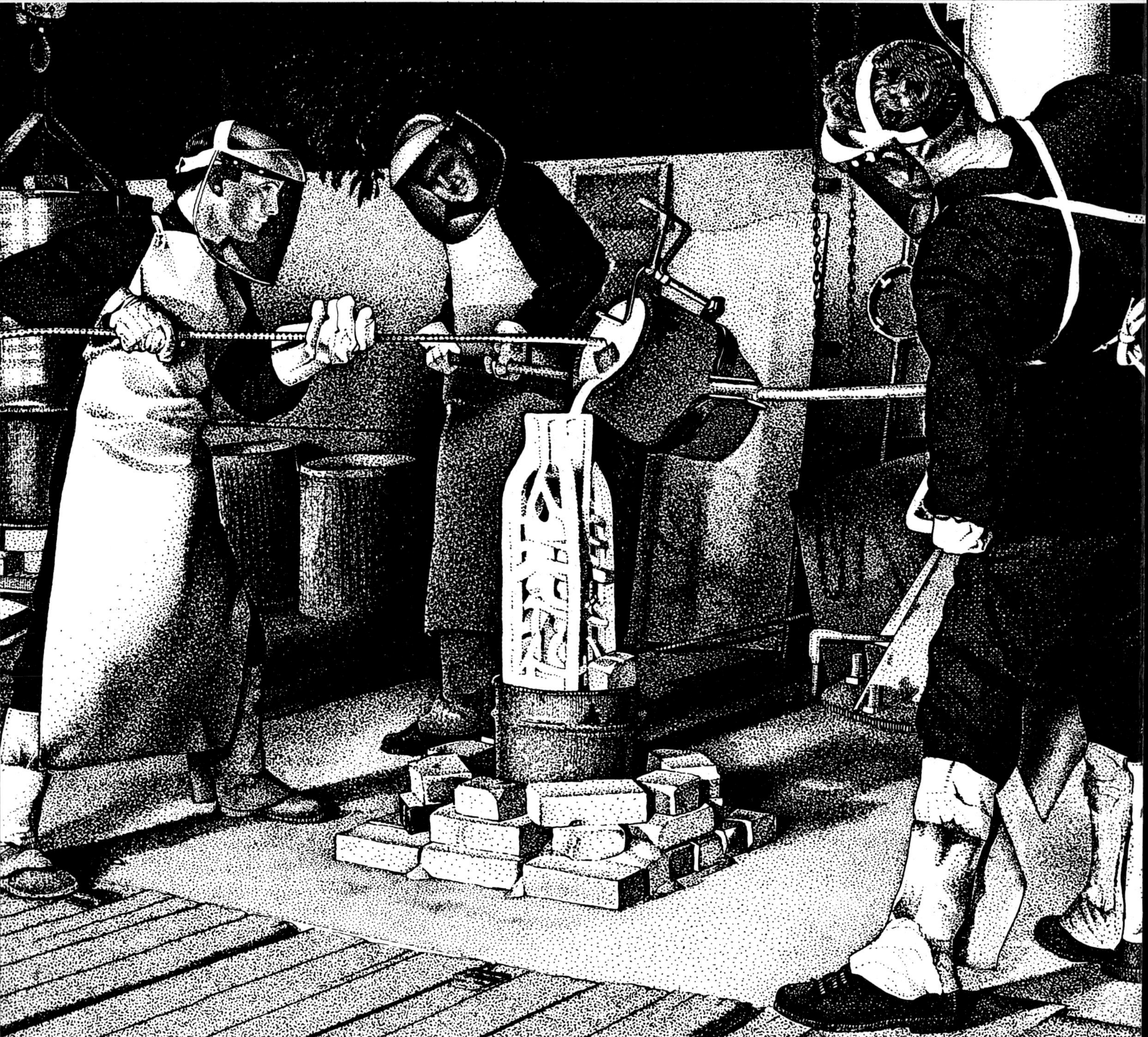

OUT OF THE FIERY FURNACE

Casting Sculpture from Ceramic Shell Molds

by HOWARD HITCHCOCK

Demonstration photographs by Peter Brenner
Drawings by the author

WILLIAM KAUFMANN, INC. 95 First Street Los Altos, CA 94022

It should be stressed that although every effort has been made to ensure and encourage safe use of materials, chemicals, and equipment, neither the author nor the publisher can assume responsibility for any injury or damage that might occur from use of the procedures described in this book.

Printed in the United States of America

10 9 8 7 6 5 4 3 2 1

Library of Congress Cataloging-in-Publication Data

Hitchcock, Howard, 1927–
Out of the fiery furnace.

Bibliography: p.
Includes index.
1. Precision casting. 2. Sculpture—Technique.
I. Title.
TS233.H56 1985 731.4′56 85-14836
ISBN 0-86576-094-2

ACKNOWLEDGMENTS

To Gerry Daniel
In memoriam

I would like to express my gratitude to the following individuals who have assisted and encouraged me in writing this book:

Betty Edwards and Donald Dame, my colleagues at California State University, Long Beach, for their generous support and continuing encouragement;

Clare Graham, Larry Davis, and David Heckaman, whose able assistance made the demonstration burnout and pour possible;

Peter Brenner, for his excellent photographs which not only illustrate the process well but also have served as the basis for most of my drawings;

Archie Boston, Jr., for his graphic design assistance on the initial book proposal, and the late Gerry Frye, who helped me get underway with the mechanicals for the book;

George and Rae Turnbull, whose design experience and sympathetic counsel helped when things looked darkest;

And all of my students, whose creative aspirations, risks, and failures have contributed greatly to our mutual understanding of the ceramic shell process.

CONTENTS

PREFACE

Lost-wax casting is truly an ancient and venerable art that has been practiced in diverse cultures and continents over thousands of years. The most modern approach to that ancient process, dating from only the mid-point of the twentieth century, is ceramic shell casting. Instead of being encased in the traditional solid-block mold that weakens in the heat of burnout, the wax model is thinly coated with a ceramic material that is fired by the burnout into a hard, lightweight shell for receiving the molten bronze. In both cases, whatever was wax becomes bronze.

Ceramic shell casting is the most spectacular art process imaginable, particularly at the crescendo of a night pour when the glowing, golden stream of liquid bronze flows into the red-hot shell, bringing it to neon incandescence in the dark. Use of this new ceramic mold material requires a distinctly different approach to various stages of the lost-wax casting process.

Out of the Fiery Furnace is the result of a twenty-year search for the simplest and most effective way to cast sculpture from ceramic shell molds with minimum cost. I have attempted in the first section of this book to make the process so clear through a sequence of photographs and drawings that anyone unfamiliar with ceramic shell casting can readily understand its complexity and sense its beauty. The second section of the book, written expressly for those who may also wish to utilize this process themselves, is distilled from two decades of experience casting all of my own sculpture from ceramic shell molds in what has been essentially a one-person studio operation, using equipment primarily designed and constructed by myself at low cost.

This extended quest began from a chance encounter in 1964 with a student who had learned about ceramic shell casting at Berkeley. Enrolled in the summer session here at California State University, Long Beach, he attempted unsuccessfully to introduce the process into a sculpture course. Although that summer class was never able to keep the slurry from settling and hardening, a problem which led the instructor and students to give up in disgust, I was fascinated by the process and its potential. It was fortuitous that only two months before, I had received my first small grant from the campus research committee for investigating simplified techniques for bronze casting. With this $365 grant I managed to buy enough materials to assemble a mixer and prepare my first batch of slurry, using only a mimeographed industrial manual as a guide. Thus began an intriguing trial-and-error adventure that has continued right up to today.

With so little technical information available, the first phase of my search was focused on acquiring enough knowledge of the process so that I could consistently produce in bronze what I created in wax. Several years of personal experimentation and improvisation led to the development of an efficient burnout/preheating system and the construction of the necessary equipment. This groundwork enabled me to introduce in 1971 the first course in ceramic shell casting on our campus, supplementing the course in traditional investment casting we already offered.

At that point I began searching not only for information for myself but also for published materials that my students might use as a text. No book on the process could be found. Not until later did I become aware of the discussions on ceramic shell casting which had taken place at three national sculpture conferences held at the University of Kansas in 1964, 1966, and 1968. The published proceedings of those conferences, organized by Elden C. Tefft, included articles by Jacques Schnier, Don Haskin, Frank Colson, and Thomas Walsh. These sculptor/teachers were pioneers in adapting the ceramic shell process to the needs of the studio sculptor and college-level art programs. It was indeed encouraging to find that my own independent discoveries closely paralleled theirs.

On the other hand, published descriptions of the

lost-wax process using traditional solid investment molds I found to be neither new nor scarce. These books, however, all perpetuate the centuries-old Renaissance approach; several even use photographic sequences shot in European foundries. They describe a process that has now been superceded in most American art foundries and higher education sculpture programs by ceramic shell casting because of its many technical advantages.

Most of what little has been published about ceramic shell casting has been written specifically for and in the technical language of industry, where the major concern is for cost-effective, highly-mechanized mass production. Broader treatises on sculptural techniques published during the past decade have referred to ceramic shell molds in only brief sections that are either too generalized or too scattered to thoroughly explain the process. Although it was not surprising to find a lack of published references on this process in 1971, books devoted to ceramic shell casting of sculpture are today still a rarity.

Next I sought to find out more about the process from commercial art foundries. In visits to major sculpture foundries from New Mexico to New York and Florida, I found to my frustration that most firms were unwilling to reveal their procedures due to fear of business competition. Visits to commercial foundries in the Los Angeles area have confirmed that technical questions usually receive only vague and generalized answers.

That was not the case, however, at the Johnson Atelier and Technical Institute of Sculpture in Princeton, New Jersey, where I had the privilege of serving as Visiting Artist during the fall of 1980. Thanks to Herk Van Tongeren and Ronald Young, Co-Executive Directors of the Institute at the time, that residency gave me the opportunity to observe a large, industrial-scale shell casting operation in detail and at length, and to compare it with the one-person, studio-scale system I had evolved on my own. The contrast was almost overwhelming—their 400-gallon stainless steel slurry tank compared to my 40-gallon resin-coated garbage can! But I was pleased to find little difference, other than the scale, between the Institute's basic approach and my own.

A student apprentice who had worked for months at the Atelier commented to me one day, "Sure, ceramic shell molds are the best way to cast, but how could I ever afford to buy such costly equipment for my own studio?" Personal experience had provided me with a ready answer to that question. I suddenly realized that besides guiding students through ceramic shell casting for the first time, I might be able to assist young artists like this apprentice who have already mastered the process and are seeking a simplified, low-cost studio method for casting their own work. College and university faculty may also be looking for an affordable way to introduce ceramic shell casting into the curriculum. This book is intended to aid in each case, since I have yet to encounter, in visiting university and commercial sculpture foundries across the country, a simpler or more efficient approach to casting small to medium-sized sculpture than that which I am presenting.

Out of the Fiery Furnace is written mainly for those who wish to make ceramic shell molds themselves. However, there are many persons who would simply like to understand how sculpture is created with this modern version of the familiar but still mysteriously intriguing lost-wax process. Comments and questions over the years from individuals viewing my own sculpture in a gallery setting indicate that most people are intensely curious about how a cast bronze is made. This book defines pictorially the multiple stages from wax to finished bronze through which a work must progress so that all readers, regardless of background, can readily comprehend how the contemporary sculptor uses a process that has evolved from traditions dating back thousands of years.

INTRODUCTION

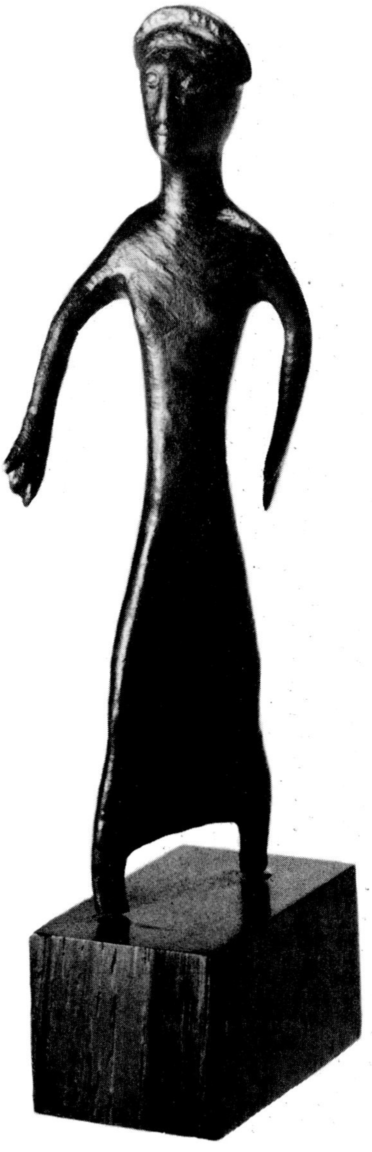

Etruscan Statuette of a Woman. Sixth Century B.C. By this era the Etruscans were well-known throughout the Mediterranean world for their bronzeworking skills. Height 11.8 cm (c.4⅝"). The J. Paul Getty Museum.

Before comparing the ceramic shell and solid-investment approaches, it should be recognized that these are simply two different ways of making molds for lost-wax casting. Thus we need first to understand the basic principle of the lost-wax or *cire-perdu* process by which a work is transformed from one material into another. In essence, a form sculptured from wax is enclosed in a heat-resistant material, a mold. The wax is then melted out, leaving a cavity into which molten bronze or other metal is poured. When the mold is broken away, the form that was originally wax has become bronze.

The discovery of this principle came very early in the history of civilization, as far back as perhaps six thousand years ago, although no one knows exactly when bronze casting began. After finding that certain materials dug from the earth—natural ores—became fluid in the fire, primitive man sought to control the shape of the molten mass as it solidified. The first molds were probably *intaglio* molds carved out of soft stone, or simply cavities made by pressing an object into dampened sand. The lost-wax process did not originate until someone modeled a form out of natural beeswax and wrapped it in clay to create a mold. Perhaps the coating was built up by painting on layers of liquid clay, a technique practiced in more recent centuries by African craftsmen. The mold was then inverted in the fire to melt out the wax and harden the clay shell. At the same time, some copper-based ore would be melting in a clay crucible for pouring into the evacuated mold. After the mold had been filled and the metal had cooled, the clay mold was broken away to reveal a replica of the original wax model. Small solid-bronze statuettes cast in this manner dating back to about 4000 B.C. have been found on the plains of northern Syria. The *Etruscan figure* to the left, made over two thousand years later, was no doubt cast by a very similar method.

From its primitive origins, bronzeworking evolved earliest in Mesopotamian cultures, and before the end of the third millenium B.C. was highly developed. The spirited *long-horned bulls* of that era on a following page attest to the influence of nomadic tribes from the north (where ore deposits were plentiful) whose invasions and wanderings dispersed a knowledge of bronze-casting techniques throughout the Middle East and beyond. As early as 2300 B.C. the Egyptians incorporated lost-wax casting with hammered sheet bronze in statues of King Pepi and his son. However, not until the Eighteenth Dynasty (after 1580 B.C.) did lost-wax casting come into widespread use. Some of the finest Egyptian bronzes, like the *cat with gold earrings* shown here, were created centuries later during the Saite period (645-525 B.C.).

Bronze is an alloy composed primarily of copper, with a smaller amount

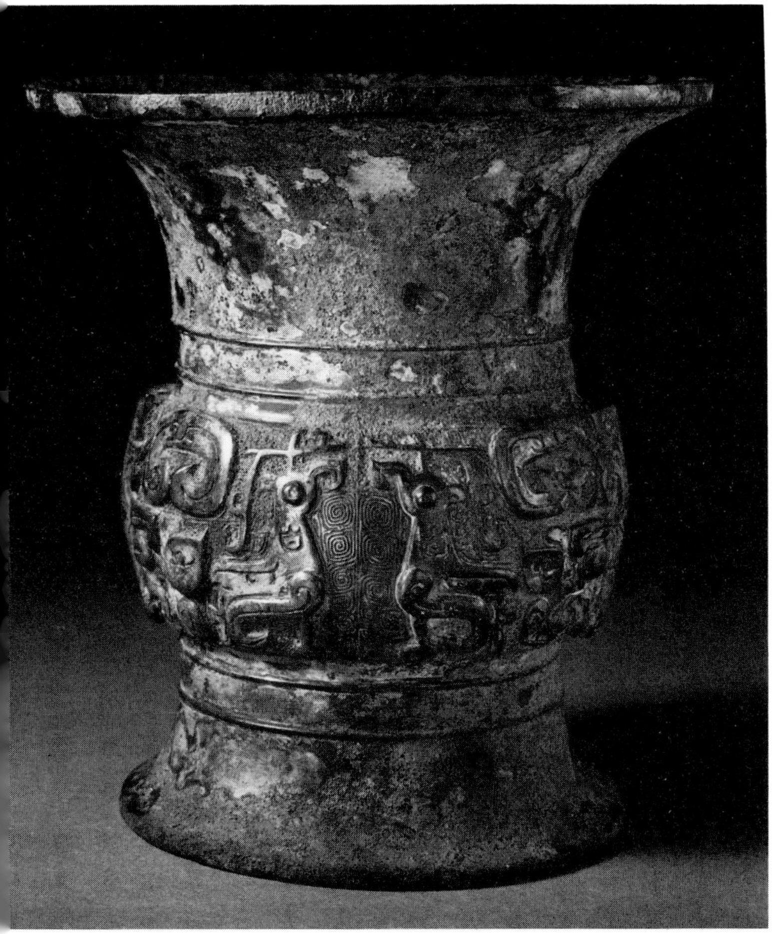

Tsun Ritual Bronze. China, early Western Chou. Tenth century B.C. Lost-wax molds were preceded in China by reusable ceramic piece-molds for casting vessels such as this. Thick slabs of clay were pressed in sections upon the carved surface of a clay model. After these fitted pieces had been dried and fired, they were reassembled around a core for pouring the bronze. Height 7¾″. The Los Angeles County Museum of Art: Los Angeles County Funds.

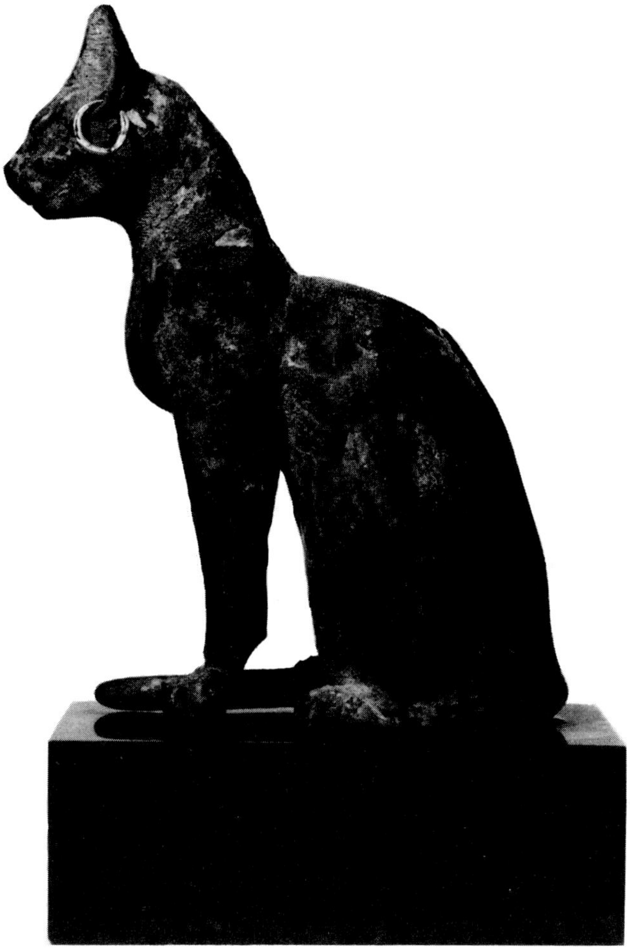

Cat. Egyptian, Saite period, c. 600 B.C. Bronze with gold earring. The cat, sacred to the goddess Bast, was a favorite pet and a popular subject for sculpture in this period. Height 6⅝″. Norton Simon Museum of Art: The Norton Simon Foundation.

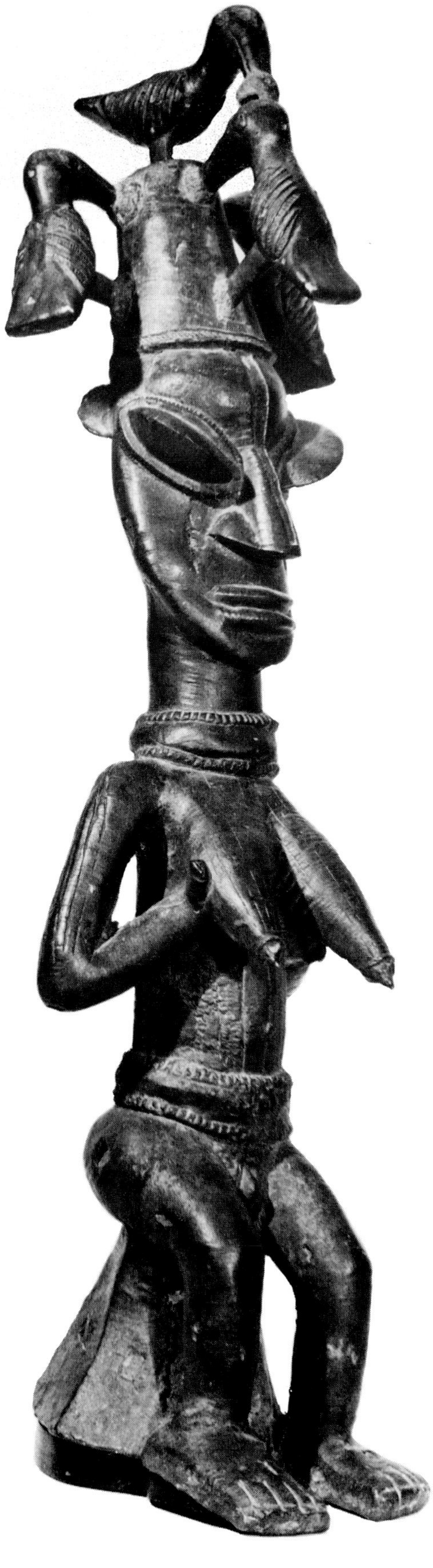

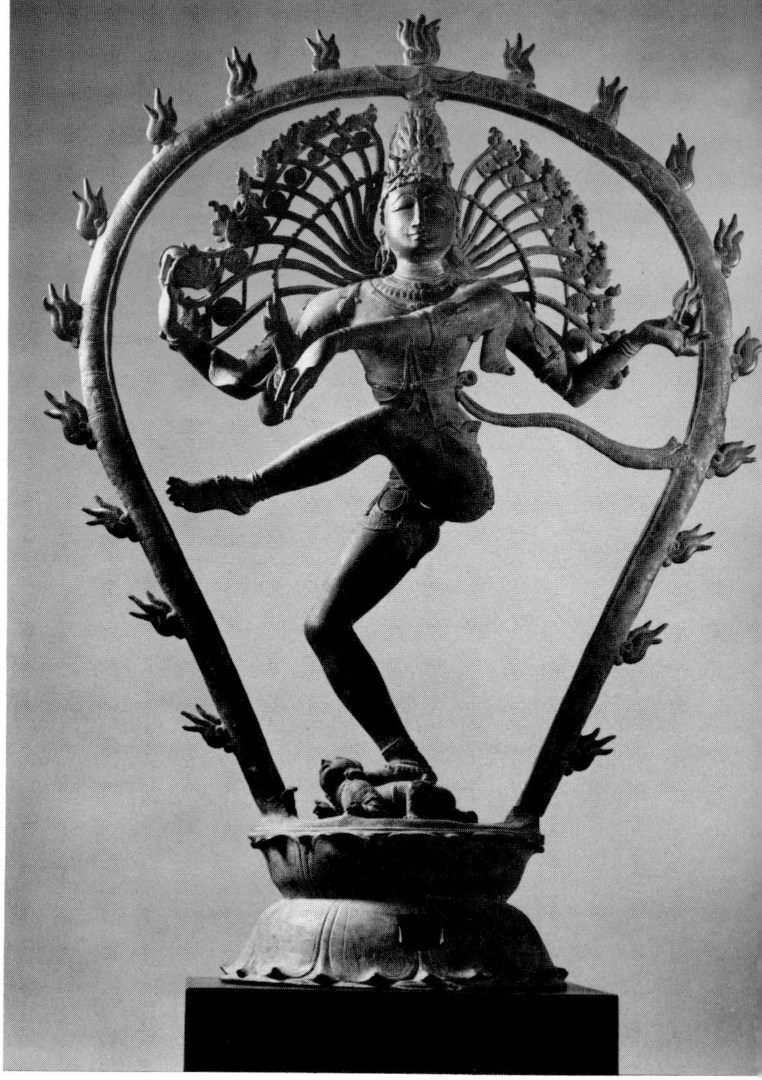

Shiva, King of the Dancers (Nataraja). India, Tamilnadu, Chola Dynasty. Tenth century A.D. The dance of Shiva personifies the cosmic forces of Nature, with the dwarf demon Ignorance crushed underfoot. Height 30″. The Los Angeles County Museum of Art: Anonymous gift.

Seated Queen. African, Nigerian, Ogboni. Seventeenth/eighteenth century A.D. In contrast to the sensitive Ife and Benin portraits, this stylized bronze seems more akin to carved wooden images. Height 32½″. Norton Simon Museum of Art: The Norton Simon Foundation.

of tin or other metals added. The most ancient bronzes must have been made from ore deposits in which copper and tin were naturally combined because the deliberate manufacture of bronze alloys did not occur much before 1800 B.C. Since copper could be readily obtained on Cyprus (from whence it got its name), the spread of bronze casting through the civilized world was dependent upon the availability of tin. Thanks to enterprising Phoenician traders who sailed as far as Britain (known then as the Tin Islands) to secure and exchange this essential metal, lost-wax bronze casting by 1500 B.C. was well-known not only throughout the Mediterrean world, but also across Europe and Asia from Scandinavia to India and China.

In China bronze fragments found in early Erlitou culture sites (1900-1600 B.C.) make clear that bronze casting had already had a long history there by that time. Ritual bronzes such as the *tsun* wine vessel illustrated were produced from ceramic piece-molds in the Shang (1573-1028 B.C.) and later periods. Lost-wax casting, however, was not used extensively in China until about the eighth century B.C.

Lost-wax casting in India dates back to the Indus Valley civilization (c. 2500-1700 B.C.) at Mohenjo-Daro and Harappa. Because tin was scarce and bronze was considered impure for casting divine figures, Indian icons are usually pure copper or brass, or have gold or silver added. The finest cast metal images, such as the *Shiva Nataraja* shown here, were produced in South India under the Cholas during the tenth century A.D. Other nations of Southeast Asia have likewise carried on a tradition of lost-wax casting over many centuries. As in India, the dominant focus has been upon creating images of the gods and

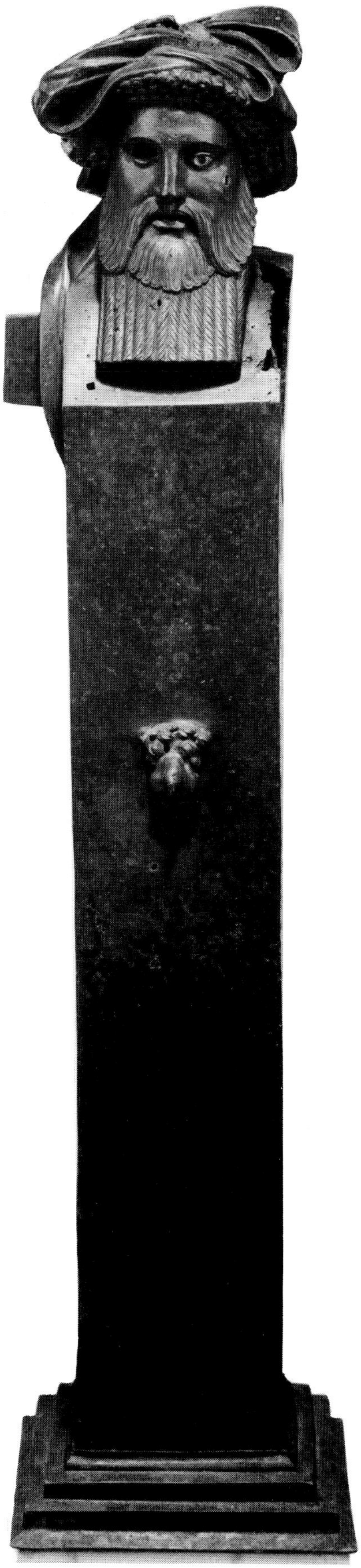

Herm, by Boethos. End of the second century B.C. Hellenistic Greek and Roman sculptors did much realistic portraiture which captured the unique personality of each subject. Height 103.5 cm (c.40¾"). The J. Paul Getty Museum.

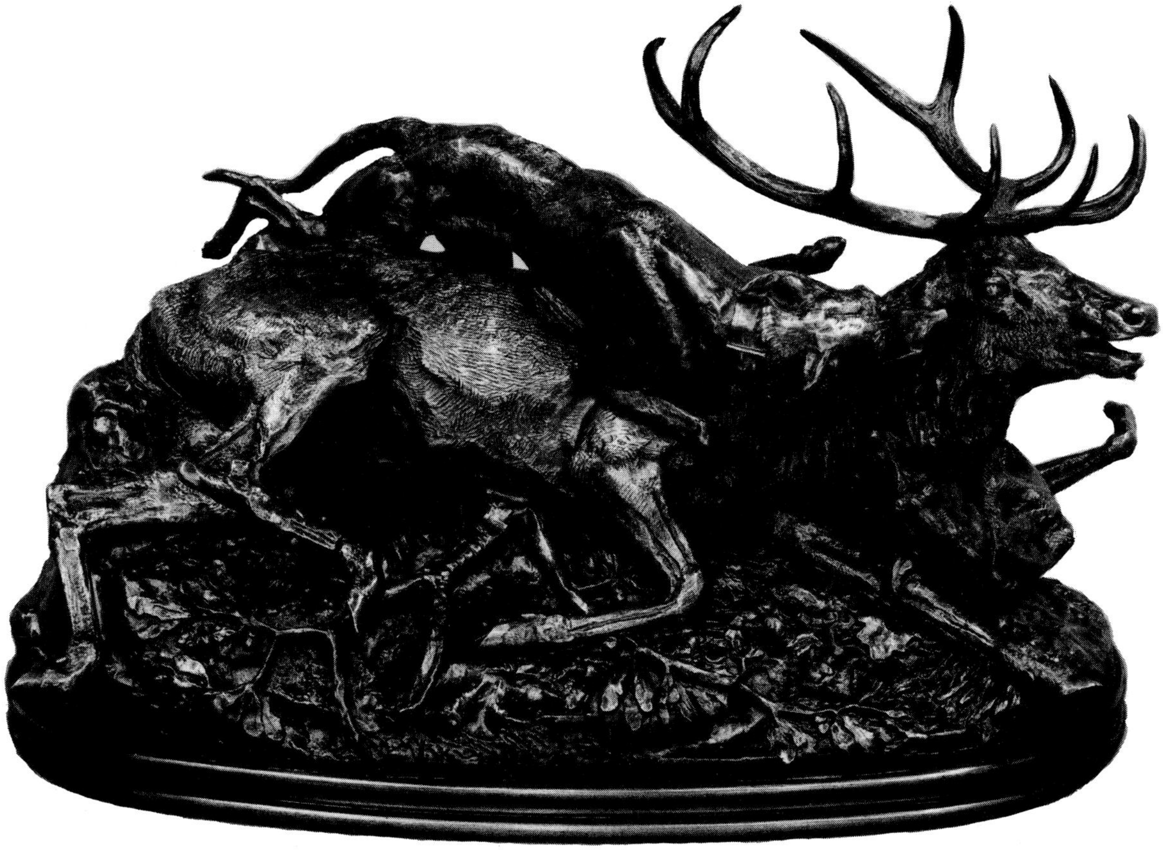

Stag Brought Down by Scotch Hounds. c. 1850, by Antoine Louis Barye, French, 1796-1875. Highly skilled in interpreting animals in combat, Barye cast much of his sculpture in his own foundry. Height 15", width 21". The Los Angeles County Museum of Art: Gift of Mrs. Leona Cantor.

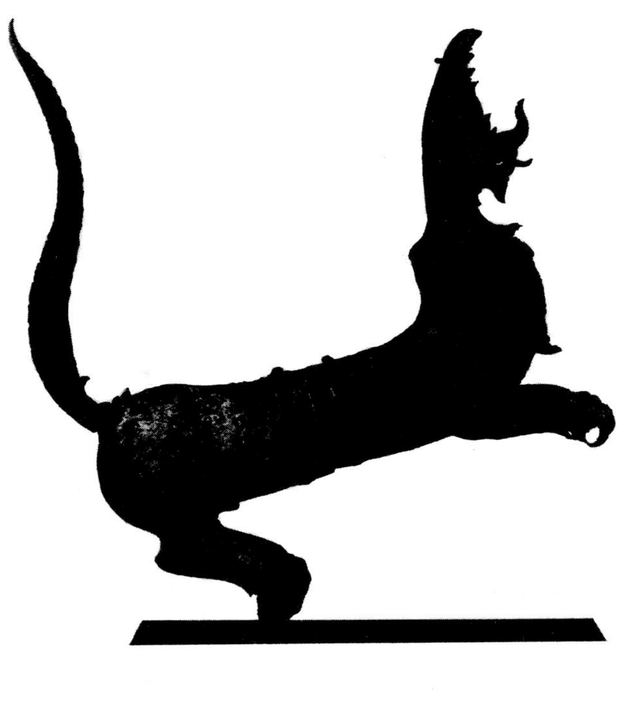

Mythical Animal (Makara). Cambodian, Khmer Period, Angkor Vat. Twelfth century A.D. This stylized and ferocious creature could have been intended to frighten away evil spirits from the temple entrance. Height 16¾", length 15". Norton Simon Museum of Art: Norton Simon Art Foundation.

enhancing temples for their worship—the *mythical animal* below on the left may have guarded the Cambodian temple at Angkor Vat.

In the Western world, the Greeks by 700 B.C. had devised a method for casting large, hollow bronzes instead of only small, solid ones. By taking thin wax impressions from plaster piece-molds made from an original model, sculptors of Classical Greece produced in well-organized workshops idealized bronze figures ranging from life-size to colossal scale. By the first century B.C. bronze figurines were being produced and marketed in great quantities, and a much greater realism crept in, as the *Herm* by Boethus illustrates. During the early centuries of the Christian era, the Romans developed factory production lines and sophisticated joining techniques (soldering, welding, brazing, riveting) in order to mass-produce armor and weapons for their legions and utilitarian objects for their households. Roman portraiture and equestrian statues were to be admired and emulated centuries later by Renaissance sculptors.

For the past two thousand years lost-wax bronze casting has flourished almost continuously, if not in one culture and continent, then in another. In Africa during the eleventh to the sixteenth centuries, the Ife and Benin cultures produced sculptured royal portraits in bronze. The lost-wax process was also used in later African cultures for more highly stylized interpretations of rulers like the *Seated Queen* illustrated on the preceding page.

Lost-wax bronze casting in Europe was strongly revitalized by the Renaissance in sixteenth-century Italy. In Benvenuto Cellini's *Autobiography* we have a colorful account of the sculptor's trials with both process and patron. After the Classical restraint of the High Renaissance typified by Michelangelo, sentiment and movement in sculpture increased, as the *Triton* of Mannerist sculptor Giovanni Bologna reveals. These tendencies peaked during the seventeenth century in the restless Baroque bronzes of Bernini, and eventually subsided into Neo-Classicism. In the nineteenth century Antoine-Louis Barye dared to set a new direction by creating vigorous animal sculptures such as the *Stag Brought Down by Scottish Hounds* above on the left. He was also an early teacher of Auguste Rodin, who ushered lost-wax casting into the twentieth century with his powerful, figurative bronzes.

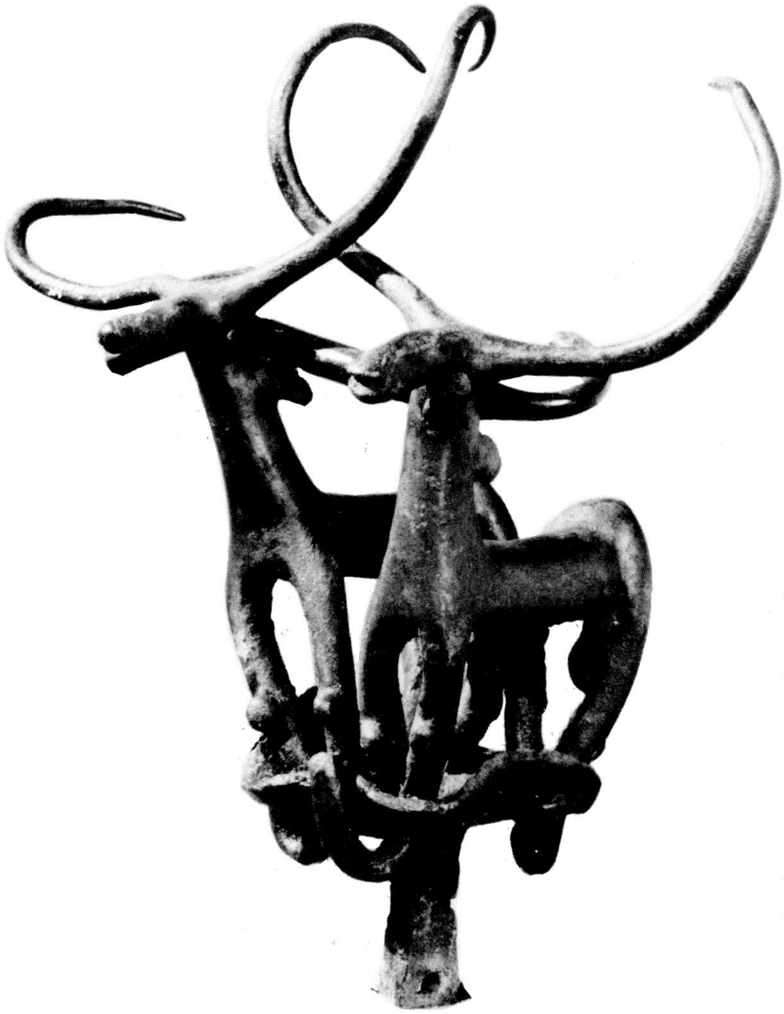

Standard, with two long-horned bulls. North Central Anatolia, near Horoztepe. Early Bronze Age, 2300-2000 B.C. Arsenical copper. Nomadic peoples in what is now Turkey depicted the animals they knew so well with fantasy and spirit. The Metropolitan Museum of Art, Purchase, Joseph Pulitzer Bequest, 1955. (55.137.5)

The examples included here are intended to represent the tremendous timespan and the geographically widespread practice of lost-wax casting. In considering the nature of the mold material, we think of ceramic shell molds as a recent technical achievement, but the very first molds, created from natural clays by unknown primitive sculptors at the dawn of history, might also be classified as ceramic shell molds. Anyone who has worked with clay, however, can imagine the inherent difficulty these sculptors faced. As a large, solid-clay mold dries, it shrinks—sometimes as much as ten percent—and severe cracking occurs. It is no wonder that alternative materials were sought, and at some point in the evolution of lost-wax casting, plaster mixed with refractory ingredients replaced clay for making molds. The ancient Greeks used gypsum to make piece-molds for duplicating their original sculptures in wax. We don't know when plaster superceded clay for molds that were to receive metal, but at least since the Renaissance, plaster of Paris has been the principal bonding agent for lost-wax sculpture molds.

During these intervening centuries no better binder could be found—until ceramic shell molds were developed following World War II. Scientists had discovered in the 1930's a method for removing most of the sodium from sodium silicate, producing a liquid colloidal silica sol that made an exceptionally strong refractory binder. A compatible dry refractory with a very low coefficient of thermal expansion was developed from fused silica; with the liquid it forms a slurry which is essentially a liquid clay. The demand for precision-cast aircraft parts that needed no machining led the casting industry to perfect technically in the 1950's the same lost-wax process that sculptors have employed with little change since the Renaissance. By practically eliminating clay shrinkage, the ceramic shell process allows us to return to the original mold material used by the earliest sculptors thousands of years ago.

MAJOR ADVANTAGES OF CERAMIC SHELL MOLDS

1. **Higher fidelity of surface reproduction.** The fired ceramic shell mold with an inner surface of fine fused silica is much harder and stronger than face coatings used in solid-investment molds. Result: reproduction of wax pattern surfaces in metal with greater fidelity, less distortion, and less shrinkage.
2. **Greatly reduced mold size and weight.** The ceramic shell mold is only ¼″ thick, thus is lightweight and only slightly larger than the wax pattern. Solid-investment molds must have walls several inches thick, a solid core, and be reinforced with wire mesh. A wax pattern coated with a ceramic shell mold, weighing altogether under 20 pounds, would require a solid-investment mold weighing 300 pounds or more! Result: greater ease in handling molds—no fork-lift or hoist required.
3. **Greatly reduced burnout time.** Ceramic shell molds are burned out in a preheated kiln within 45 *minutes;* solid molds must be heated and cooled very gradually, which takes 2 to 6 *days.* In addition, about 90 percent of the wax from ceramic shell molds is reclaimed, while all wax in solid molds is consumed and passes into the atmosphere. Result: major savings in fuel and wax replacement costs; great reduction in air pollution.
4. **No cores needed for hollow castings.** A thin ceramic shell can be built both inside and outside of hollow wax patterns at the same time, whereas solid-investment molds required a solid, vented core held in place with core pins. Result: a much lighter mold with no core pin holes to repair.
5. **No backup required during metal pour.** Ceramic shell molds need only to be propped upright for pouring, while solid-investment molds must be securely packed in dampened sand. Result: much time and energy saved between burnout and pour; no flasks needed.

6. **Can be repaired and reheated.** After burnout, fired ceramic shell molds can be reinforced or repaired if cracked or broken, redipped, refired, or stored indefinitely before pouring. If solid molds crack, however, little can be done to reinforce them; if not poured at the proper point in cooling, they are too fragile to be reheated. Result: molds salvaged that would otherwise be lost.

7. **More rapid cooling and mold removal.** When poured, ceramic shell molds cool rapidly and can usually be quenched within 20 minutes; much of the mold falls away as the metal cools and shrinks. Solid molds, in contrast, must cool for an hour or more before the castings can be broken out. Result: quicker and easier mold removal, finer grain structure in the metal, less mold waste to dispose of.

To illustrate the process of casting sculpture from ceramic shell molds, I have used an original work of mine called *God Save the Prince II,* created directly in wax. The burnout, pouring, and patination were done in the Art Department foundry at California State University, Long Beach, while all of the other stages were accomplished in my studio nearby. This work was chosen to challenge the process and to demonstrate how complex a sculpture can be cast in a single piece. I hope that the mystery and fascination of this modern version of the ancient lost-wax process will be revealed to you in the pages that follow.

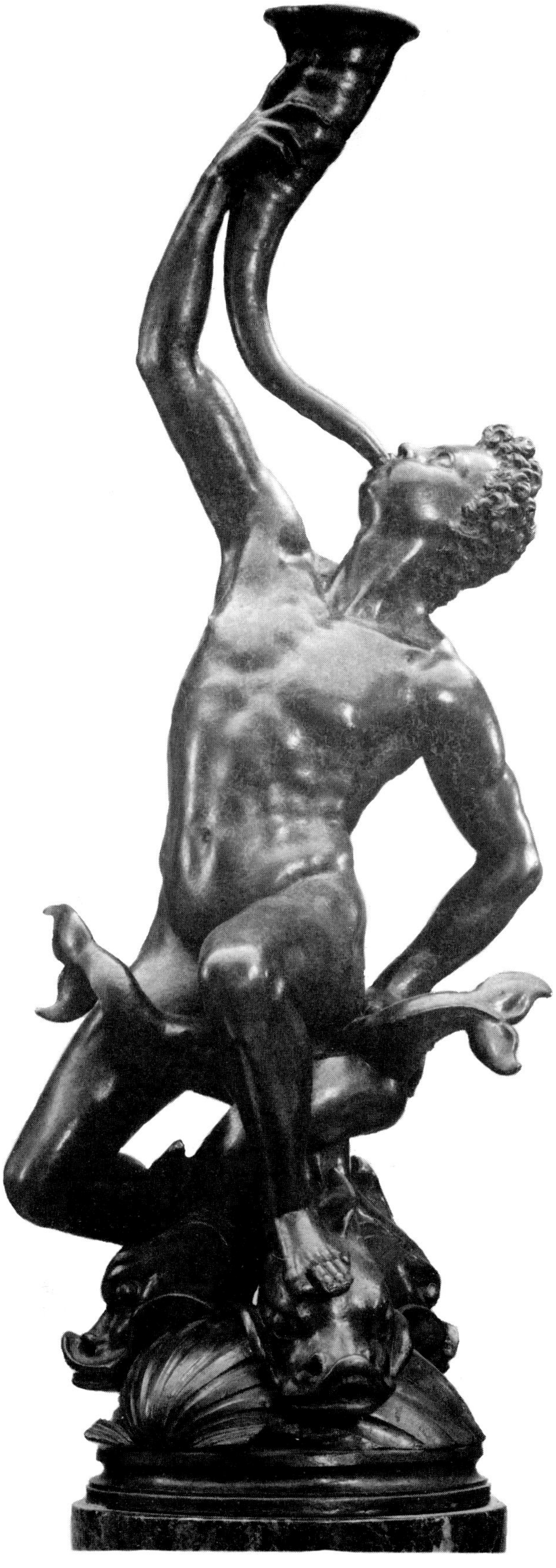

Triton. 1560-70, by Giovanni Bologna, Italian, 1529-1608. After the strongly classic style of the High Renaissance had given way to Mannerism, Giovanni Bologna became recognized as the principal Mannerist sculptor of the latter half of the sixteenth century. Height 36". The Metropolitan Museum of Art, Bequest of Benjamin Altman, 1913. (14.40.689) All rights reserved, The Metropolitan Museum of Art.

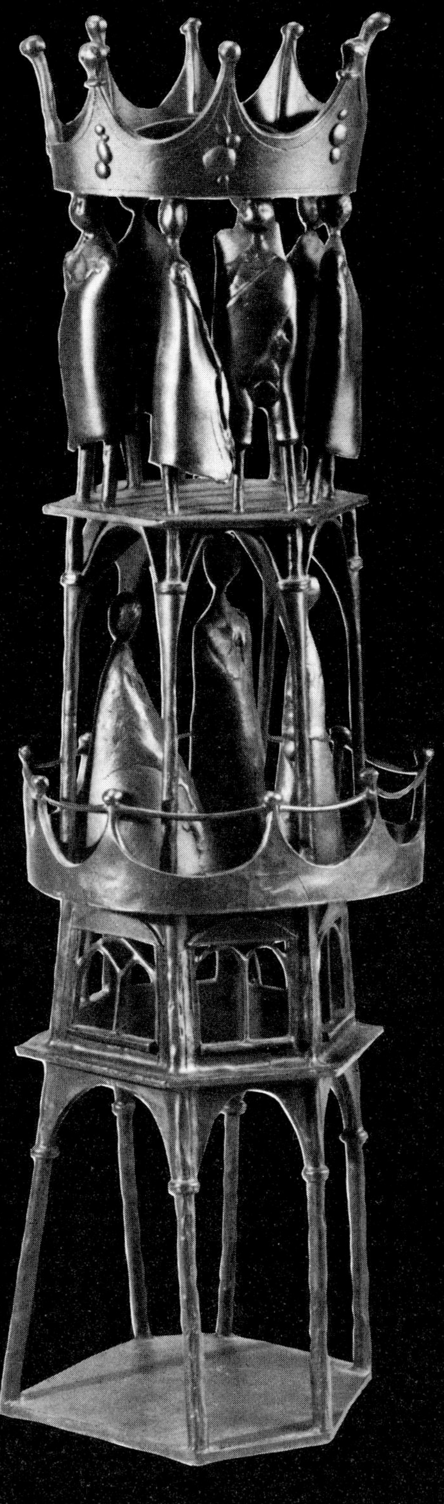

PART ONE

THE CERAMIC SHELL PROCESS

A Visual Definition

1 CREATING THE ORIGINAL WAX SCULPTURE

1.1 Originally done in charcoal, the sketch of the sculpture to be created is enlarged by the grid method to a full-scale ink drawing. The dimensions for each section of the sculpture can now be measured directly from the larger drawing. More often, however, I omit this step, and enlarge the piece as I construct it directly in wax from a thumbnail sketch no more than one or two inches in size. The maximum size for the sculpture was determined by the interior dimensions of the burnout/preheat kiln and the slurry container.

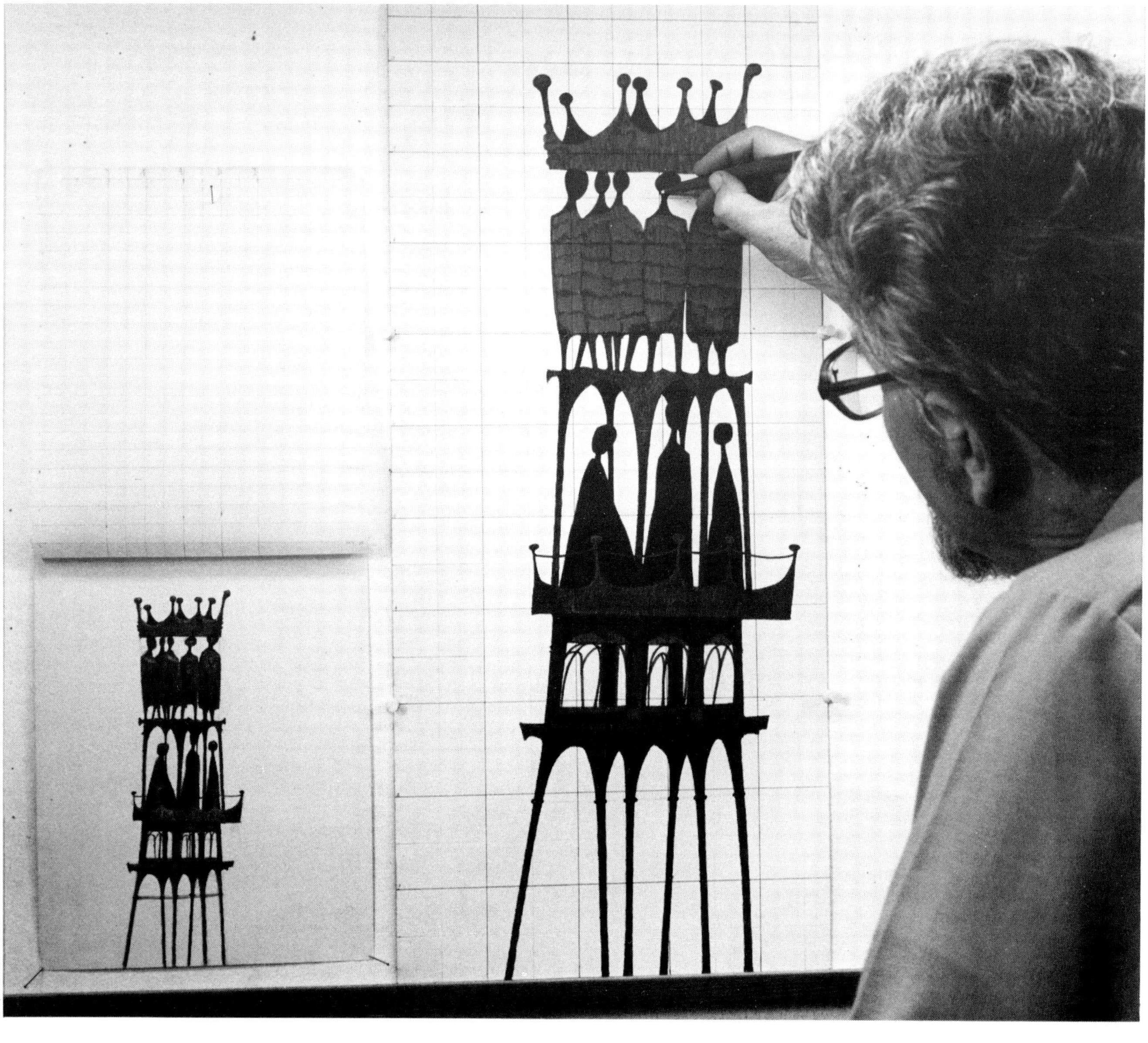

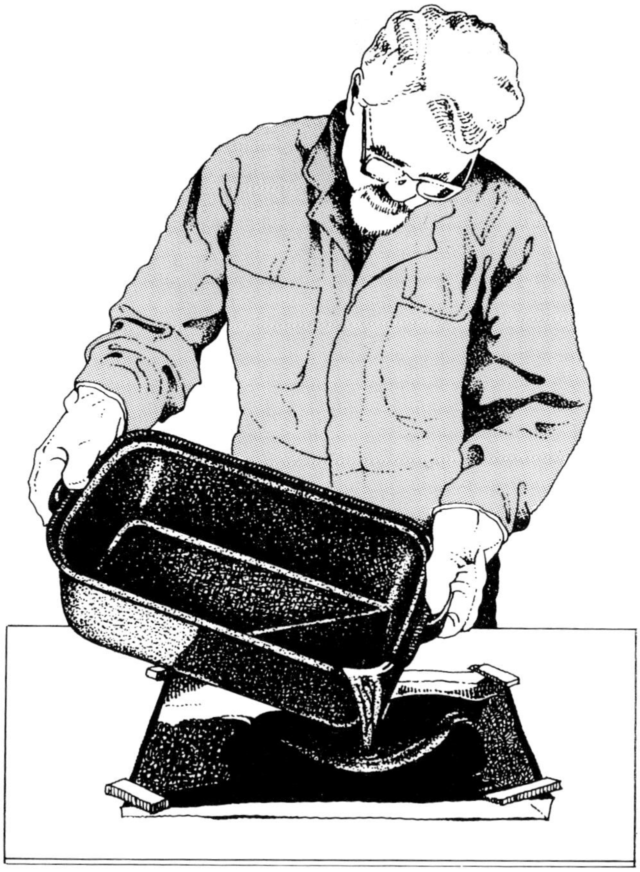

1.2 Molten wax is poured into a form made of heavy-duty aluminum foil to create wax sheets approximately 3/16″ thick. The wax is predominantly microcrystalline, with paraffine added for hardness and powdered household dye for color. The foil should be peeled off before the wax cools completely.

1.3 Using paper patterns developed from the full-scale drawing, the flat horizontal elements of the sculpture are cut from wax sheets with a sharp knife and steel ruler. Similar patterns will be prepared for cutting the vertical windowed panels and the balcony railing.

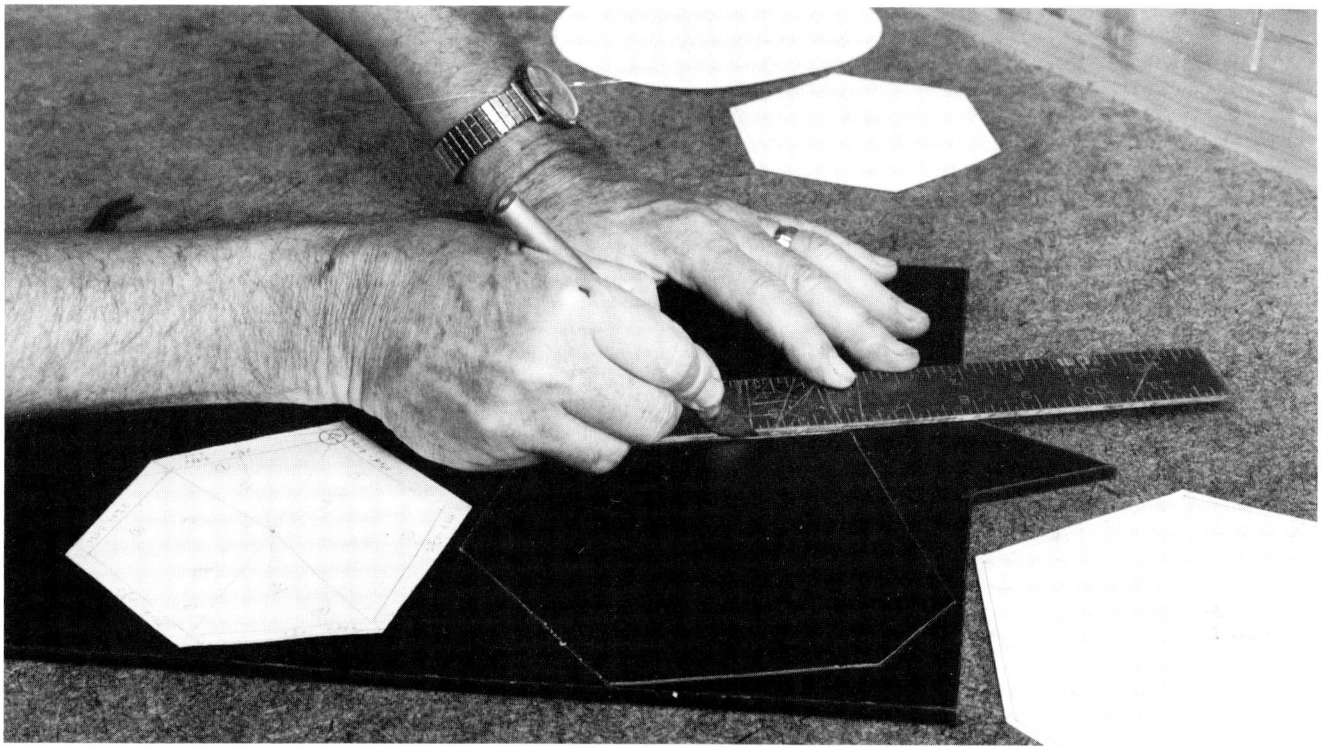

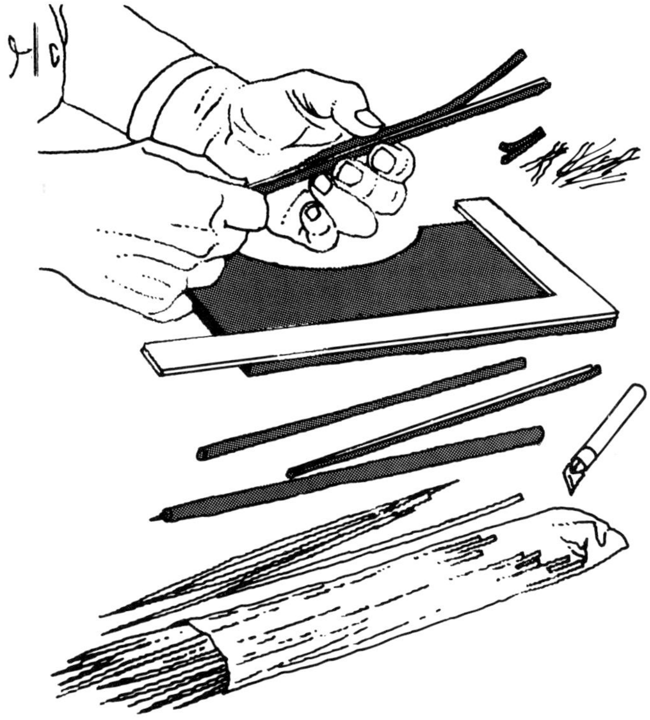

1.4 Bamboo skewers are used to reinforce the vertical columns. The skewer is sandwiched between two narrow strips of wax which are squeezed together to embed it in the center of the post. Toothpicks will be used in a similar fashion to strengthen the slender legs and necks of the figures.

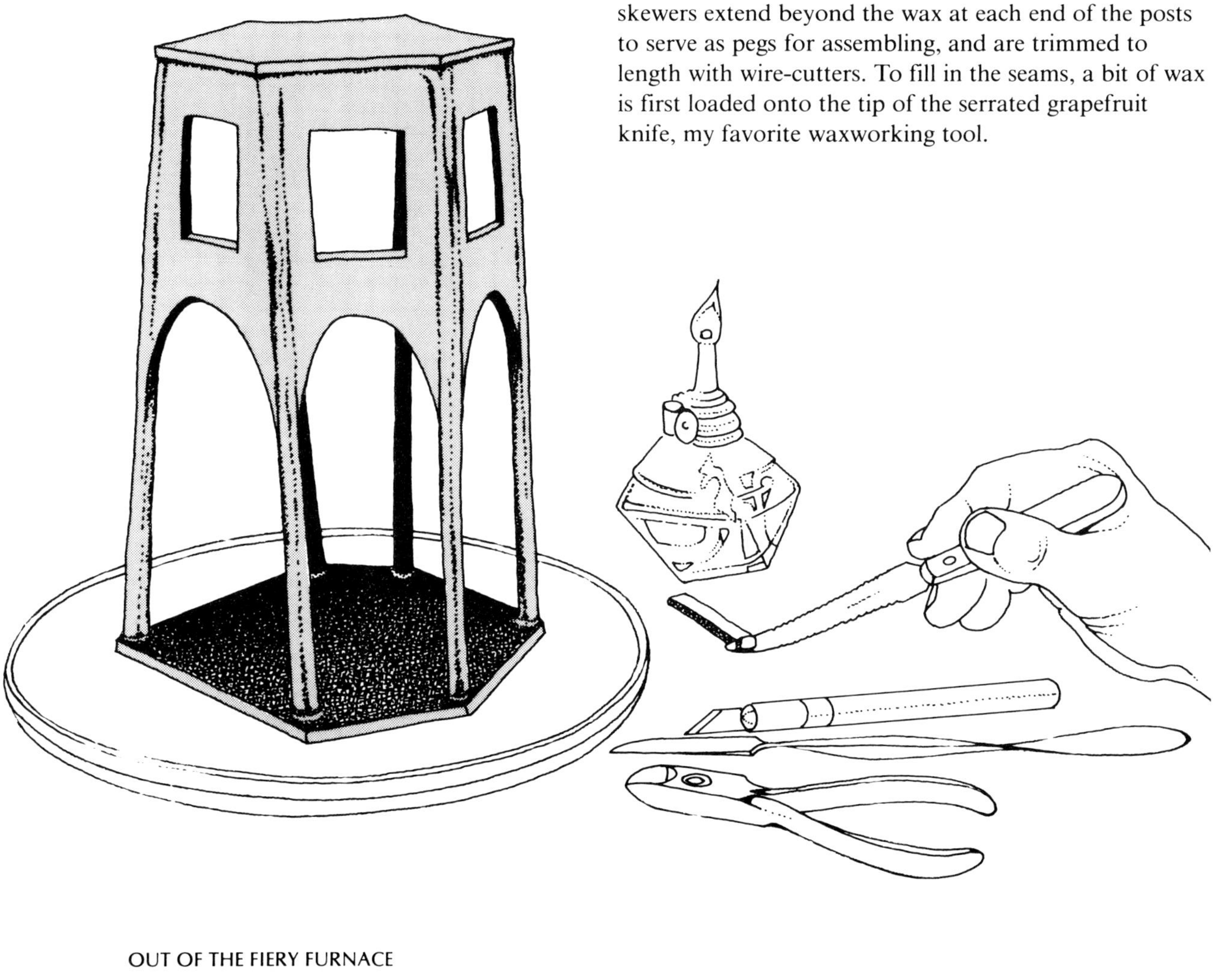

1.5 The precut panels and modeled columns are now joined by pressing the wax firmly together. The bamboo skewers extend beyond the wax at each end of the posts to serve as pegs for assembling, and are trimmed to length with wire-cutters. To fill in the seams, a bit of wax is first loaded onto the tip of the serrated grapefruit knife, my favorite waxworking tool.

1.6 The wax on the tip of the knife blade is melted in the flame of an alcohol lamp. The curvature of the blade and the surface tension hold the wax on the knife. The lamp, available at a jeweler's supply house, burns denatured alcohol fuel which can be found at most paint or hardware stores.

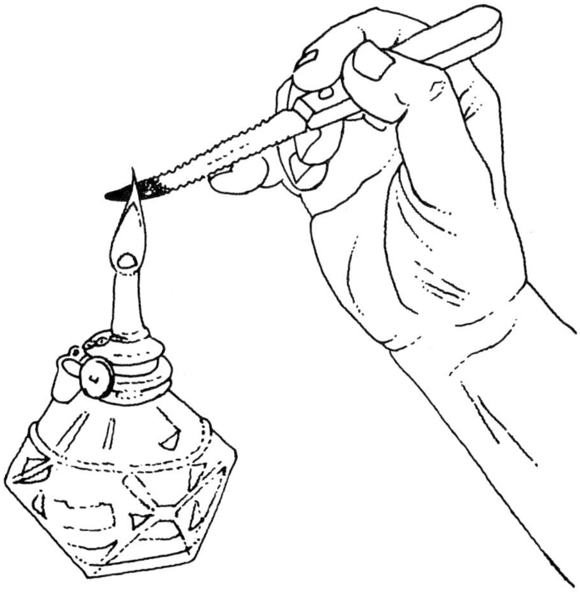

1.7 The hot wax flows from the knife tip into the joint, forming a solid and secure weld between the wax components. Since the sculpture will be handled a great deal in the construction, sprueing, and dipping stages, it is essential that all parts be joined as solidly as possible. The steel modeling tool, which can be heated in the lamp flame if desired, is useful for scraping and burnishing wax surfaces.

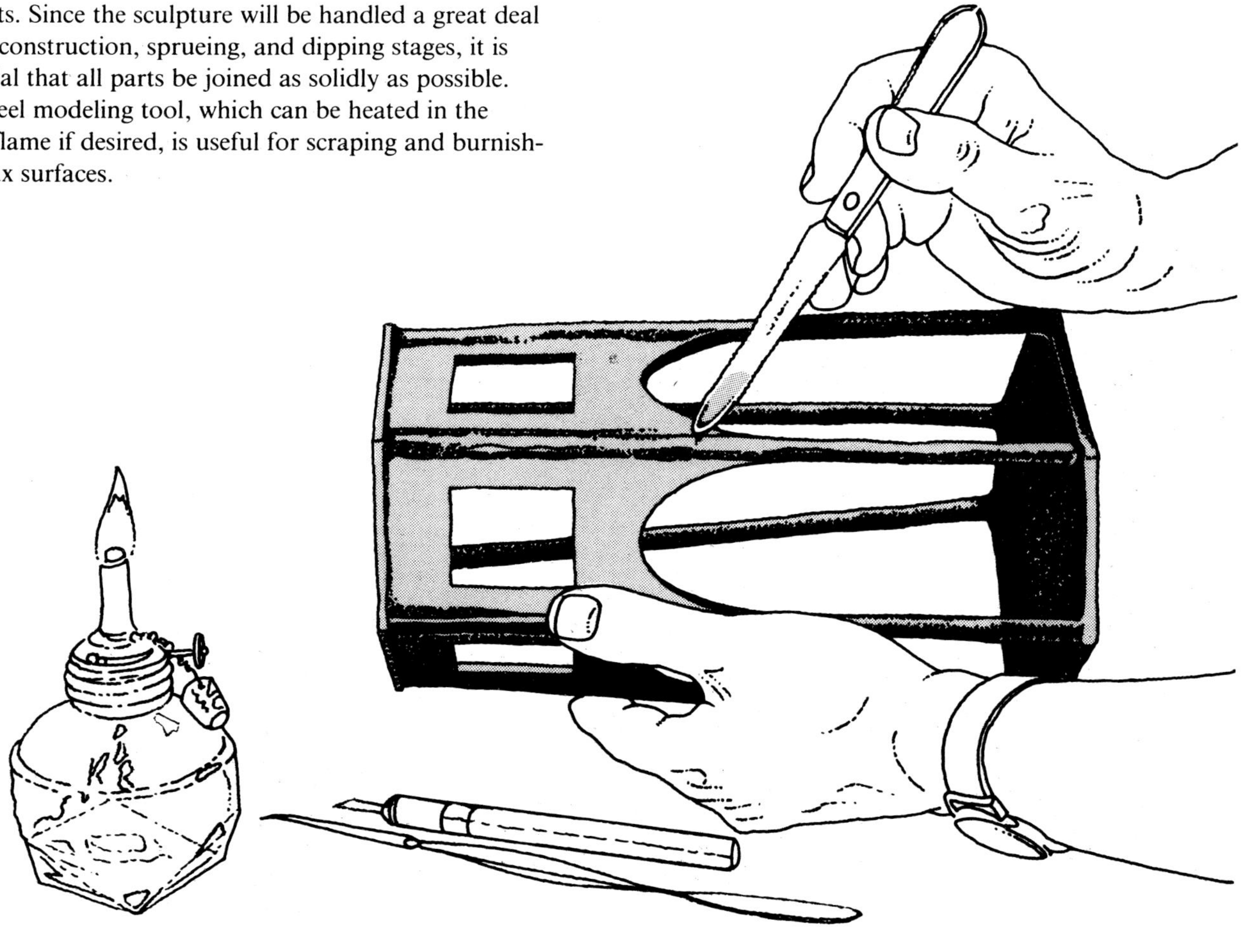

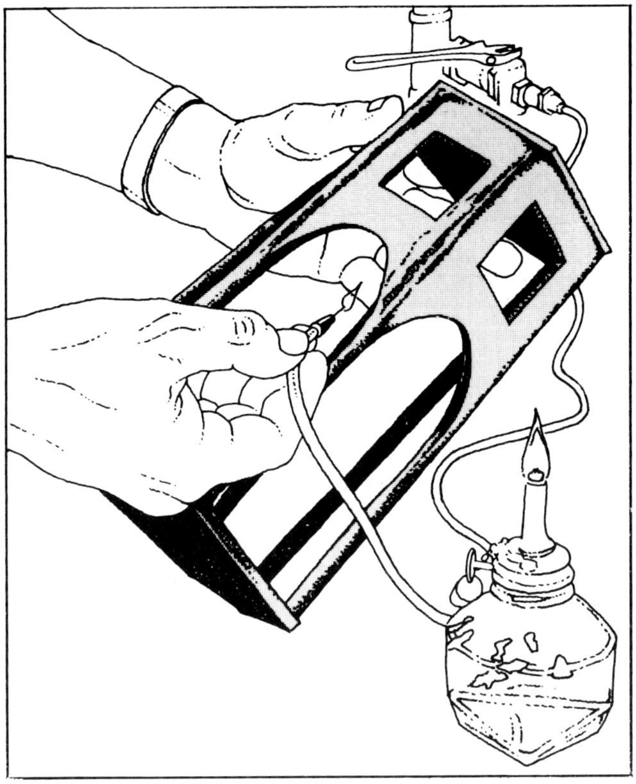

1.8 Functioning exactly as a welding torch does on metal, this tiny gas torch fuses the deposited wax into a smooth and even seam. Improvised from a hospital intravenous needle, the torch requires a sensitive adjustment; it can be relit immediately from the alcohol lamp if it goes out.

1.9 After a paper pattern had been designed for the arches of the second-level ceiling, each piece was cut from wax sheet and bent to the proper curvature. By melting the edges on an electric pancake griddle set at low heat, each arch is trued to fit exactly with the rest. With the columns added, the total assembly will be inverted and placed atop the previously-built structure.

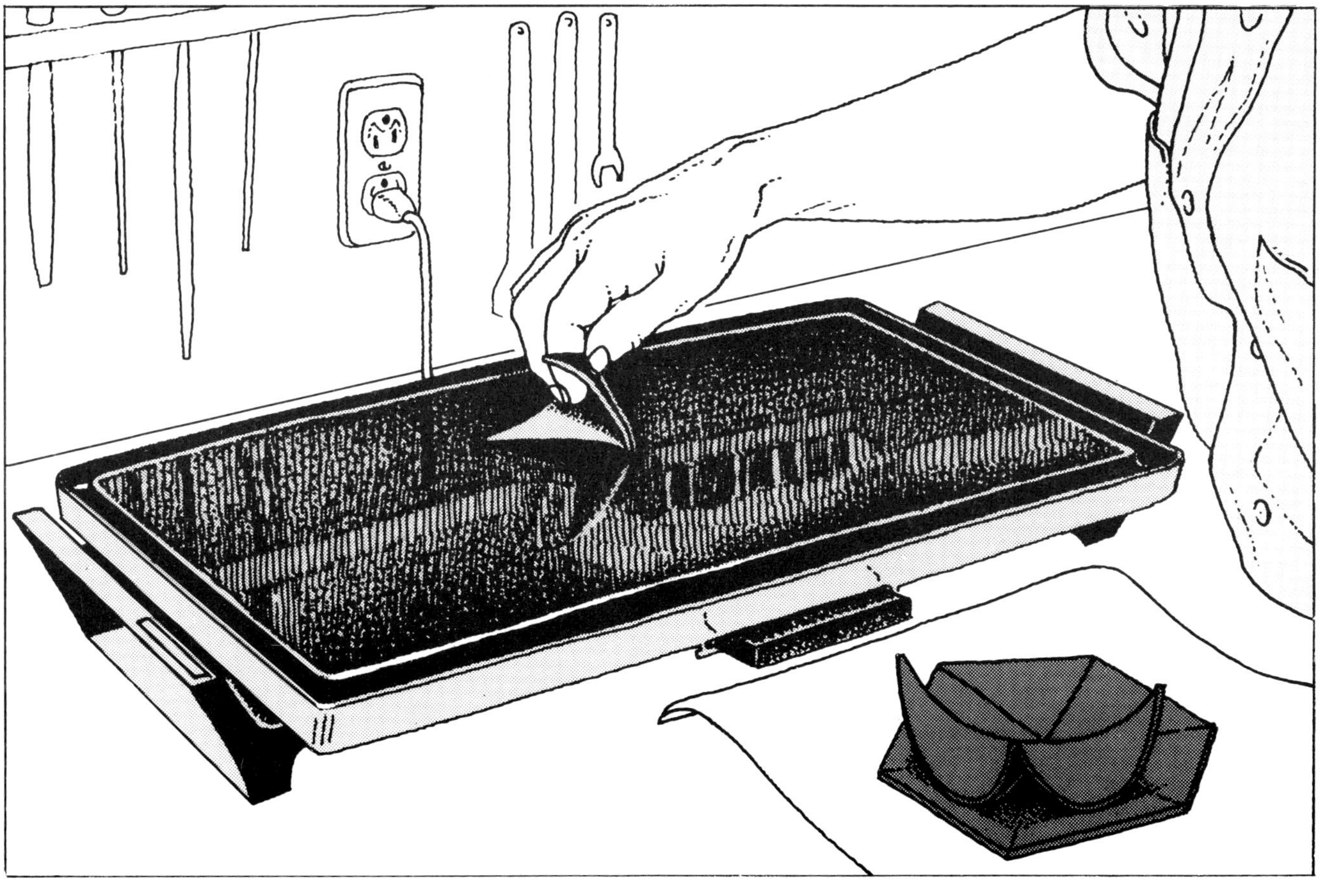

The rest of the sculpture is also constructed directly out of wax using similar techniques. Each robed figure is begun by tearing off a piece of sheet wax, then shaping it into a cone or cylinder. Into the top of this volume is inserted a head on a toothpick, and into the bottom, a pair of toothpick-reinforced legs. The wax is manipulated to form the neck and shoulders and to contour the body. Melted wax is poured into the inverted body interior up to about hip level, which eliminates the need for a core. The robes of the three center figures extend all the way to the floor. This allows the floor beneath each figure to be cut away, opening up the hollow body from underneath.

The crown at the top is cut from wax sheet using a paper pattern. The tips are reinforced with toothpicks, as are the balcony posts. Thinly-rolled ropes of wax suspended from those posts droop of their own weight. The final details include the delicate window mullions and the narrow bands accenting the columns and crown points.

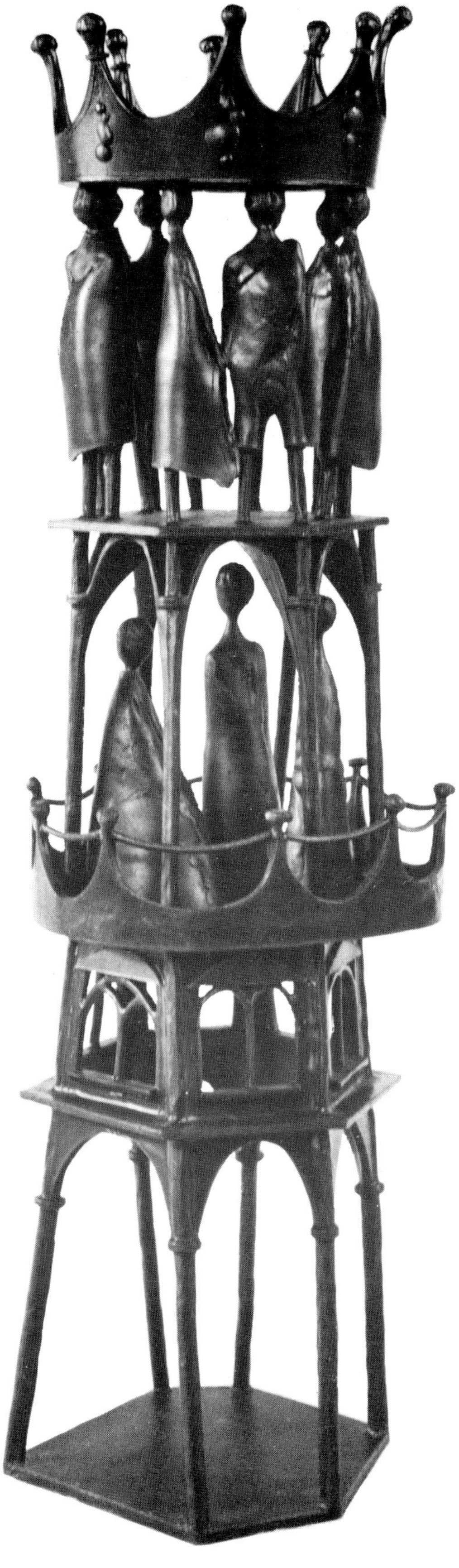

1.10 The completed wax structure stands 25″ high. The thickness ranges from about 1/16″ in the window mullions and balcony rail to ¾″ in the upper torsos of the figures. Such a wide contrast between thick and thin sections is a challenge to any casting method, and has been deliberately included here to test the effectiveness of the ceramic shell process.

2 SPRUEING AND VENTING THE WAX

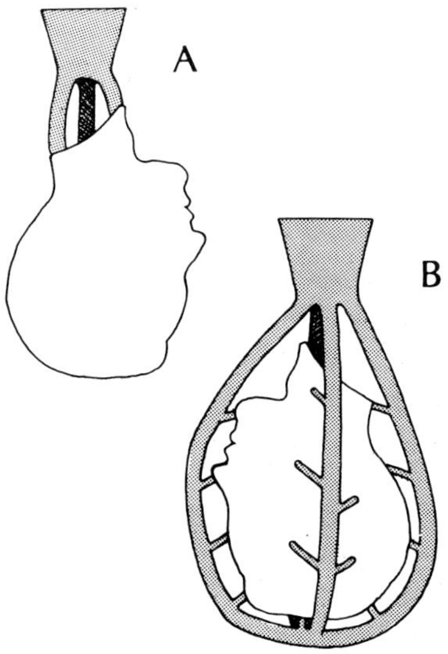

2.1 Sprueing sytems: *direct* (A) vs. *indirect* (B). A. Metal flows directly into mold cavity. B. Metal flows to bottom through main sprues, then rises to fill mold cavity indirectly.

With the wax sculpture complete, the next task is to plan a *sprueing system,* a network of hollow channels through which the molten metal can flow into the mold that will be formed around the sculpture. This input system will be created by attaching to the sculpture a configuration of wax rods, called sprues or gates, surmounted by a wax cone. After the shell mold has been built up around them, these wax rods will be·melted out, leaving hollow conduits that lead into the mold cavity. When the wax cone is melted out, it will leave atop the mold a funnel-shaped cup large enough to make a good target when pouring bronze from the crucible.

Besides its primary function of providing a means for getting metal into the mold, the sprueing system also serves two other important purposes. Its second function is to provide a drainage system, in conjunction with the vents, for evacuating wax from the mold during burnout. The same network of channels designed to permit metal to enter the upright mold will provide an escape route for molten wax to flow down and out of the inverted mold. A third function of the sprueing system unique to the ceramic shell process is to reinforce thin, fragile wax sculptures during mold construction. A surrounding structure of sprues can serve as an exoskeleton to protect the wax model from breakage or distortion during immersion in the slurry and application of stucco coatings.

The first decision that must be made in planning a sprueing system is

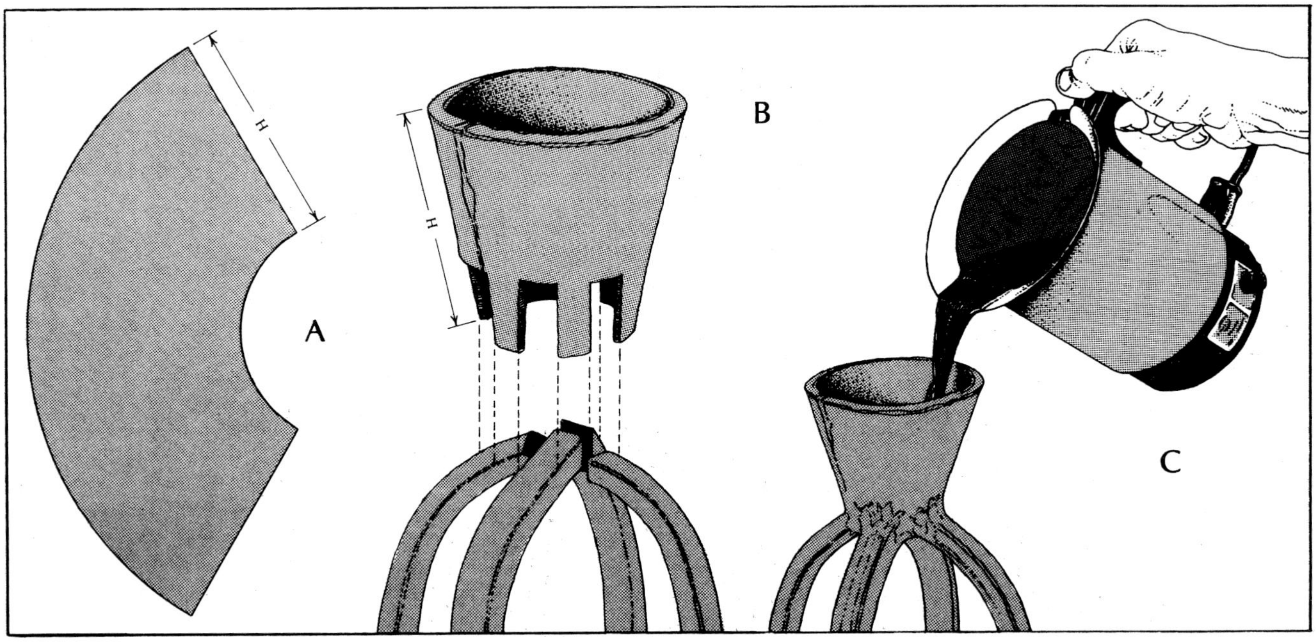

2.3 Wax runners slant upward from the main sprues to various sections of the sculpture. *Each runner must be as thick as that section of the sculpture it feeds.* Small wax rods, known as vents or risers, are attached at all upper points where air would be trapped when metal rises in the mold.

whether to use a *direct* or an *indirect* approach. From the illustration at the left, it is obvious that much less wax is needed for direct sprueing, which means less weight in both mold and casting. If the entire mold cavity can be filled from the top quickly and completely with little turbulence in the metal flow, then the direct approach is definitely better. However, the indirect approach has to be used if the interior surfaces of the mold are either (a) too irregular, which would cause the entering metal to splash and perhaps chill on the way down, or (b) too close together, creating spaces so narrow that the metal would solidify before completely filling the mold. Since both of these conditions existed in the sculpture created for this demonstration, I had no choice but to sprue it indirectly.

The number, size, and placement of the wax sprues and vents must be determined next. The main sprues must be large enough to get the metal into the mold as quickly as possible. For small to medium-sized sculpture, ¾″ solid wax rods are usually adequate, although thicker works should have heavier but hollow sprues. Often three main sprues are arranged in tripod fashion to achieve the greatest strength and stability. For our method we position the cone so that the inverted mold will balance upon it in the burnout kiln. The runners on the sculpture to the right range from ¾″ square rods feeding the solid figures to ¼″ round rods feeding the thinnest sections.

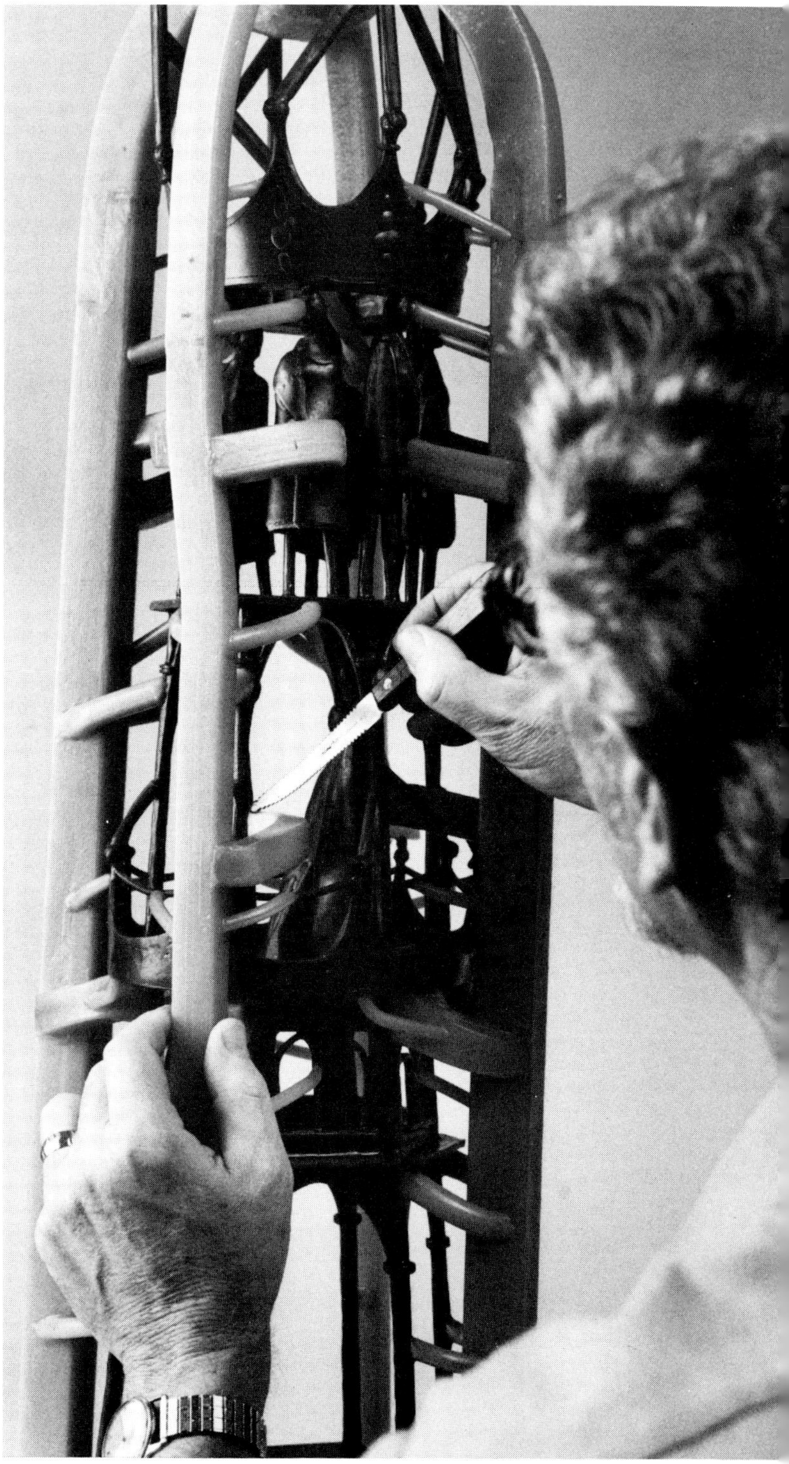

◀ 2.2 A hollow wax cone on top of the sprueing system will create a cup to receive the metal in the finished mold. From sheet wax a flat shape (A) is cut; this is then curved until the straight edges meet to form a cone (B). Notches cut into the bottom edge allow the cone to fit down over the intersecting main sprues. After the cone has been securely modeled around the main sprues, it is filled halfway with molten wax (C) to ensure a solid, strong connection.

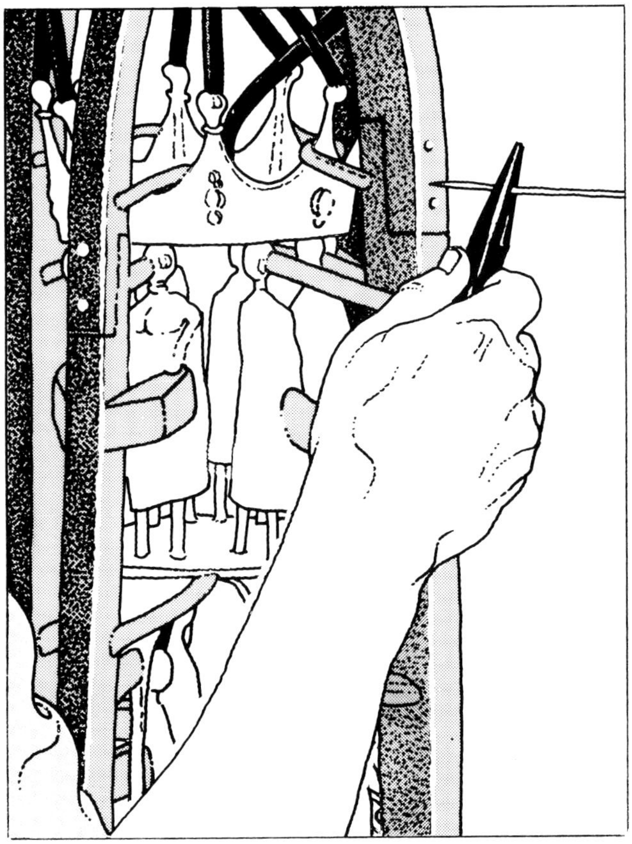

2.4 If wax rods are not long enough, extensions may be spliced on. Here a bamboo skewer is being forced through the half-lap joint in a ¾″-square main sprue to pin the joint securely. The bamboo will be snipped off flush with the surface with wirecutters before filling the joint with molten wax. When the shell is fired, these bamboo pegs will burn up completely.

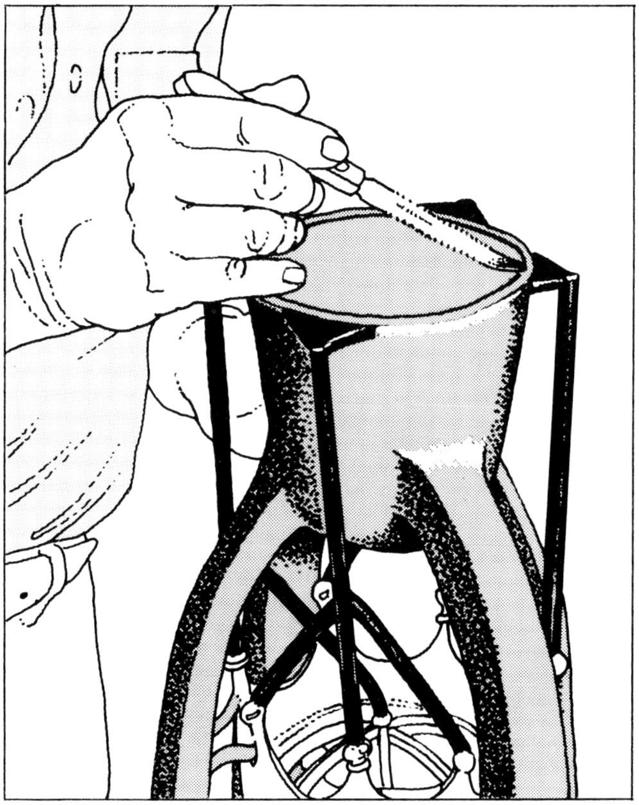

2.5 The final step in completing the sprueing is to seal the pouring cone, or cup. A circle is cut from wax sheet to fit inside and level with the top of the cone. A wax disc is being welded into place here with the heated tip of a knife. This lid will keep slurry out of the cup when the sprued wax is immersed in the slurry later.

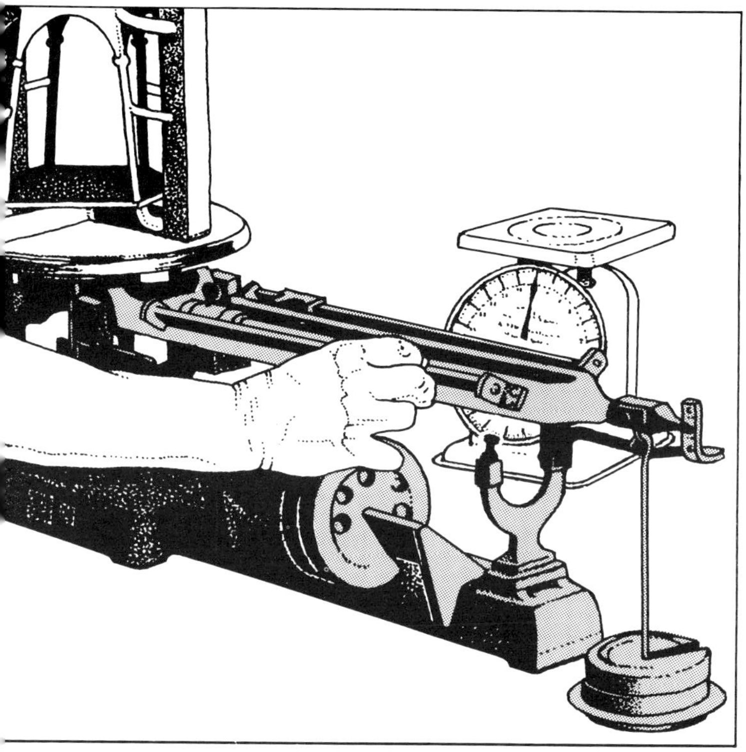

2.6 Before beginning the mold, the sprued wax must be weighed so that the amount of bronze needed to fill the mold can be calculated. For each pound of wax, ten pounds of bronze will be required. When using a gram scale as shown here, the weight in grams must first be converted into pounds (divide by 454). On a regular scale, ounces must be converted to fractions of a pound before multiplying by ten.

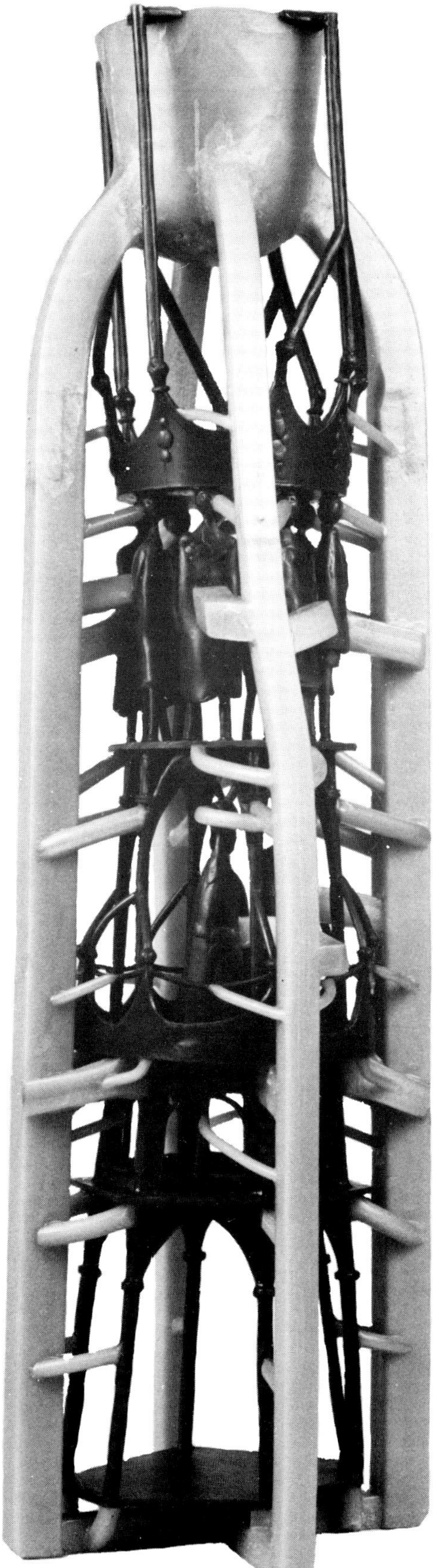

2.7 The wax sculpture is now completely sprued and vented. Note that the vents are held out away from the pouring cup by triangular spacers to keep the metal, when poured, from flowing downward into the vents. With this indirect system, the metal will flow to the bottom of the mold, then rise gently to fill the main cavity that will be the sculpture and push the air upward and out through the vents.

3 BUILDING THE CERAMIC SHELL MOLD: INVESTING

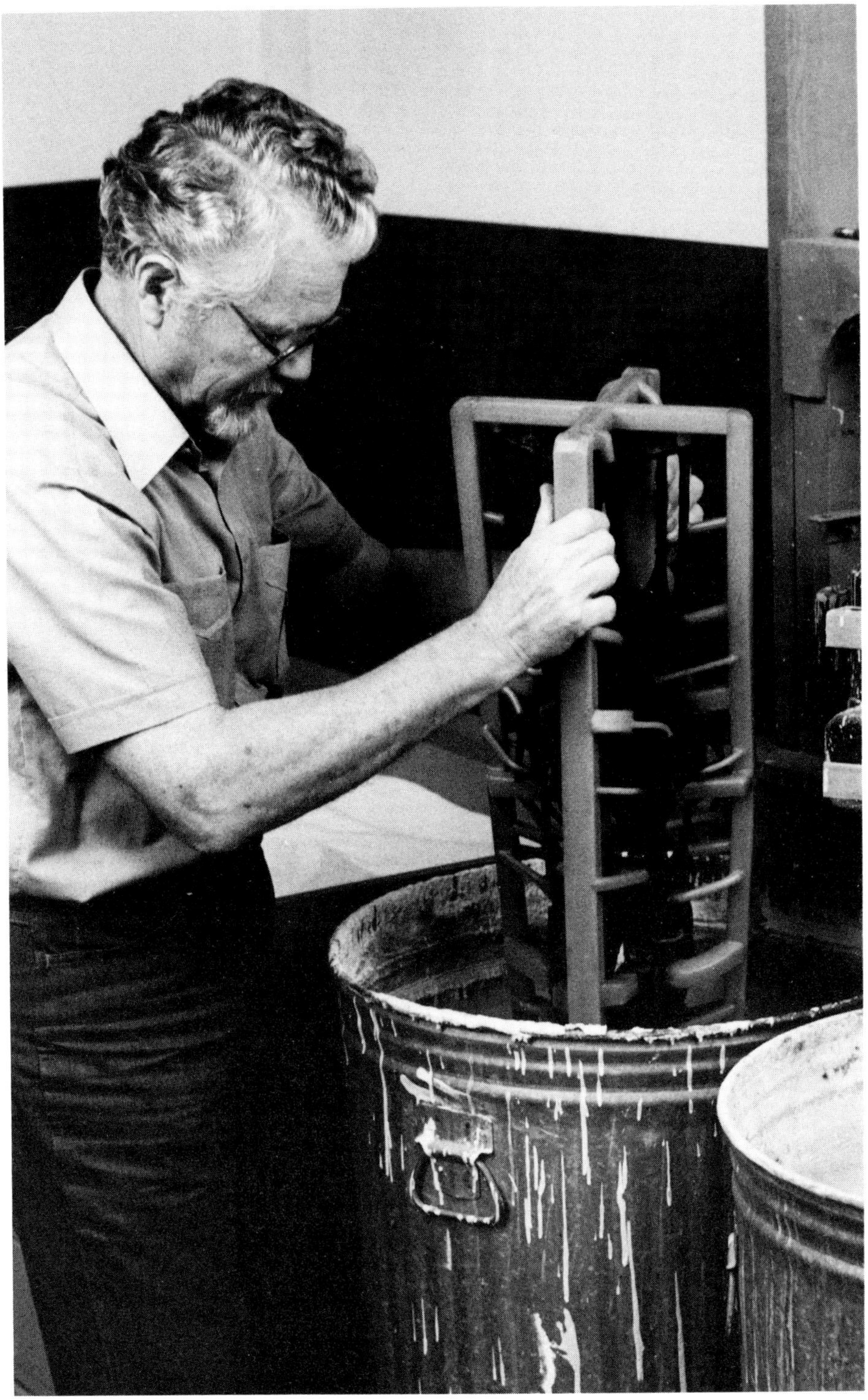

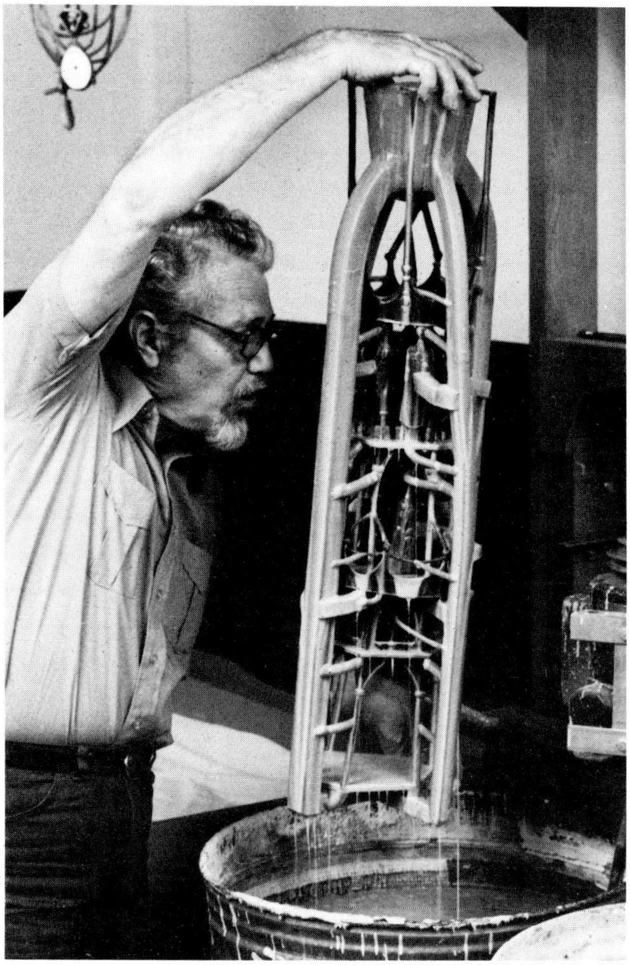

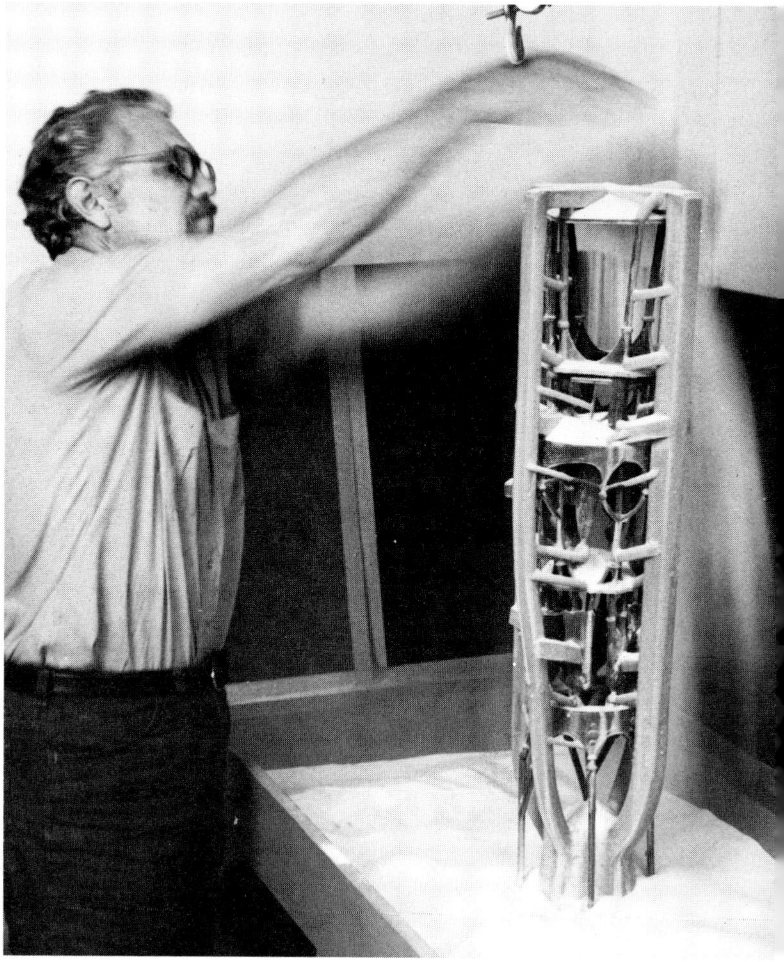

3.2 As the slurry drains off the wax, any air bubbles in corners or crevices are broken by blowing on them. The grey slurry is about the consistency of pancake batter, and must be stirred at regular intervals to keep it from settling. An automatic timer turns on the mixer for three minutes every half hour.

◀ 3.1 The ceramic shell mold is built up on the sprued wax in layers by immersing the wax repeatedly in a fused silica slurry. Each thin coating is sprinkled, or stuccoed, with fused silica sand and allowed to dry. Here the hinged mount for the mixer has been swung out of the way, and the sprued wax is being pushed gently down into the slurry.

3.3 After being dipped and drained, the wax is carefully set down in the sand. Quickly washing the slurry off my hands, I sift fine sand (stucco) over the coated wax. This first coat must be allowed to dry thoroughly. For subsequent coats, the wax is first wet in a colloidal silica liquid before it is immersed again in the slurry. Fine stucco is used for the first two or three coats, then coarse stucco for the remaining coats to build the shell thickness more rapidly. A face mask with dust filter is strongly advised for this step.

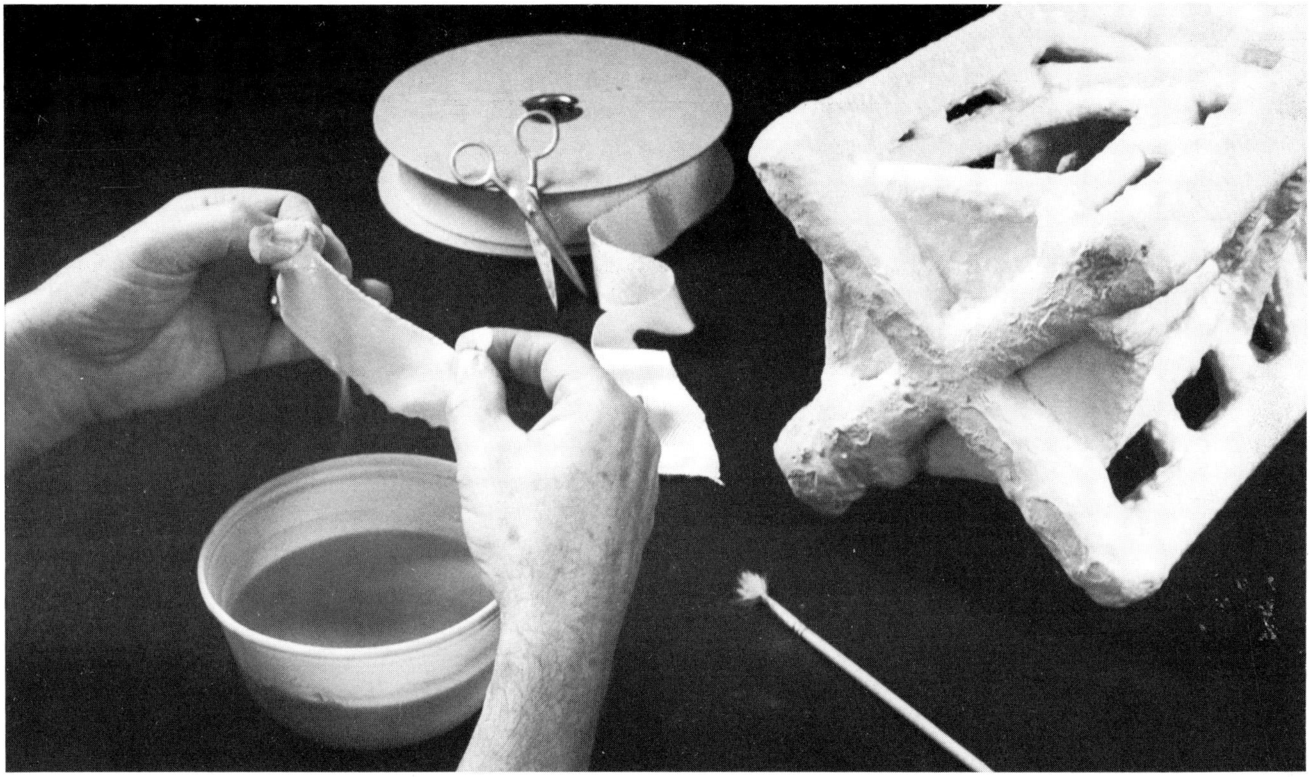

3.4 Sharp corners and narrow edges are susceptible to cracking and must be reinforced. A special absorbent, high-temperature fiberglass tape is dipped into some slurry, then applied to the edge. Fine stucco thrown on immediately thickens the slurry so the tape can be smoothed down. Applying the tape several coats before the final coat will embed it in the shell wall.

After the first coat dries completely, subsequent coats can be applied at intervals as short as one hour by using a special stucco which chemically gels the slurry coat. Maximum shell strength, however, requires thorough drying of every coat. Fans can be used to accelerate drying, but never heat, which would expand the wax and crack the shell. Both the temperature and humidity should be kept as constant as possible. The total number of coats required depends on the slurry viscosity and size of the work to be cast. Eight to ten coats are sufficient for most molds which fit within our burnout kiln. This mold actually exceeded the height of the kiln interior, and took twelve coats. No stucco is applied to the final slurry coat, which seals in loose grains of sand that might get into the mold cavity. At least twenty-four hours must be allowed for the last coat to dry before burnout.

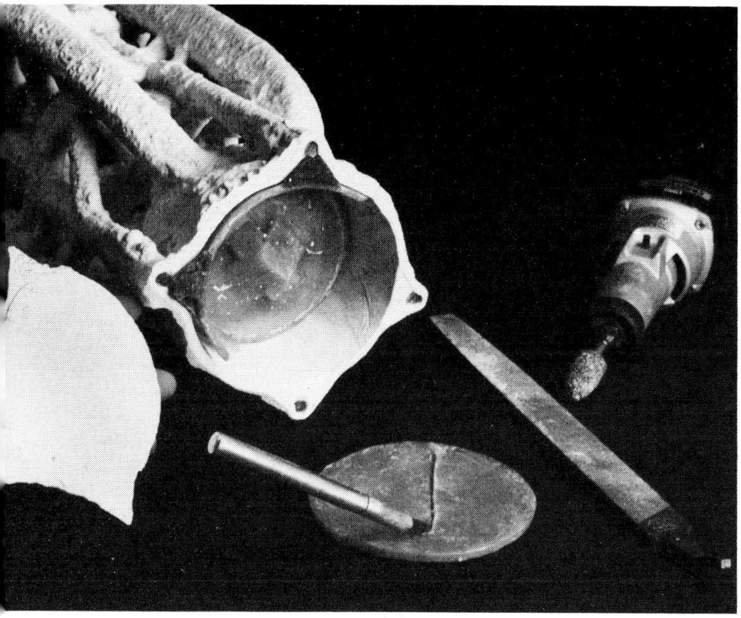

3.5 Before burnout can occur, the mold covering the top of the cone must be removed. With a hand file or grinding wheel, a bevel is cut through to the wax all around the cone. The disc of mold material is pried off, then the sheet wax lid over the cone is removed. The wax cone itself may be cut and peeled away also, down to the solid wax, if it can be done without damaging the shell.

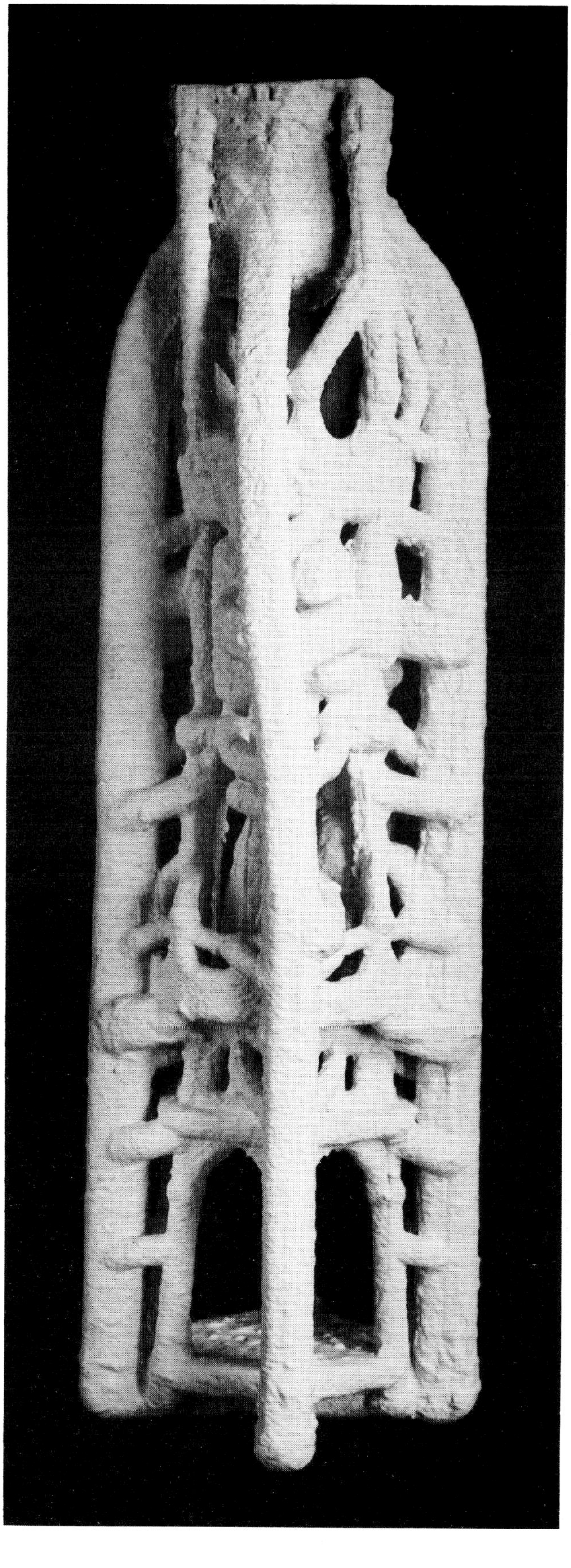

3.6 The ceramic shell mold surrounding the sprued wax has been completed and reinforced. With the top of the cone removed, the mold-coated wax is now ready for burnout. The shell averages ¼″ in thickness and adds about twenty pounds to the weight of the wax.

4 DEWAXING AND FIRING THE SHELL: THE BURNOUT

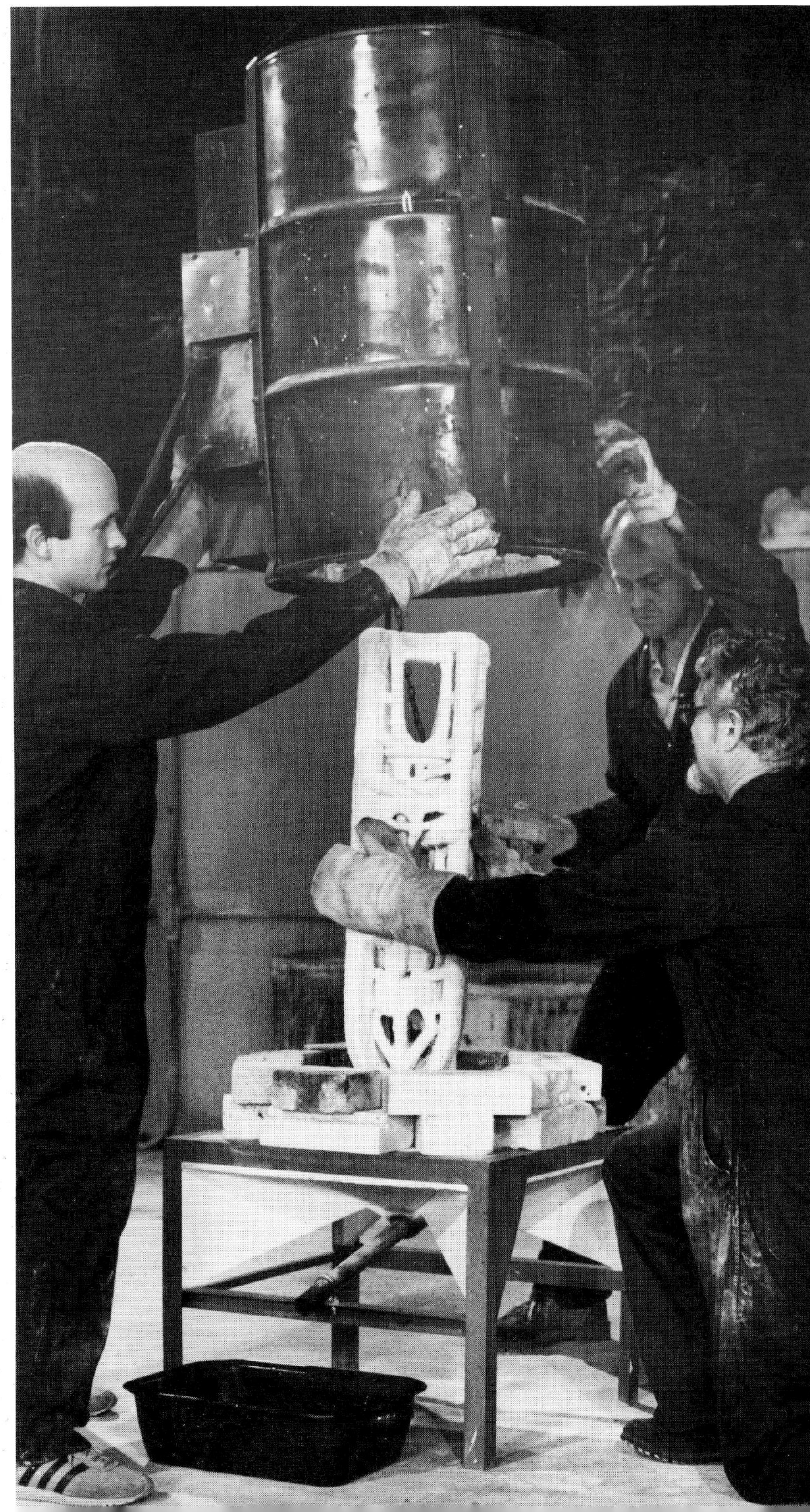

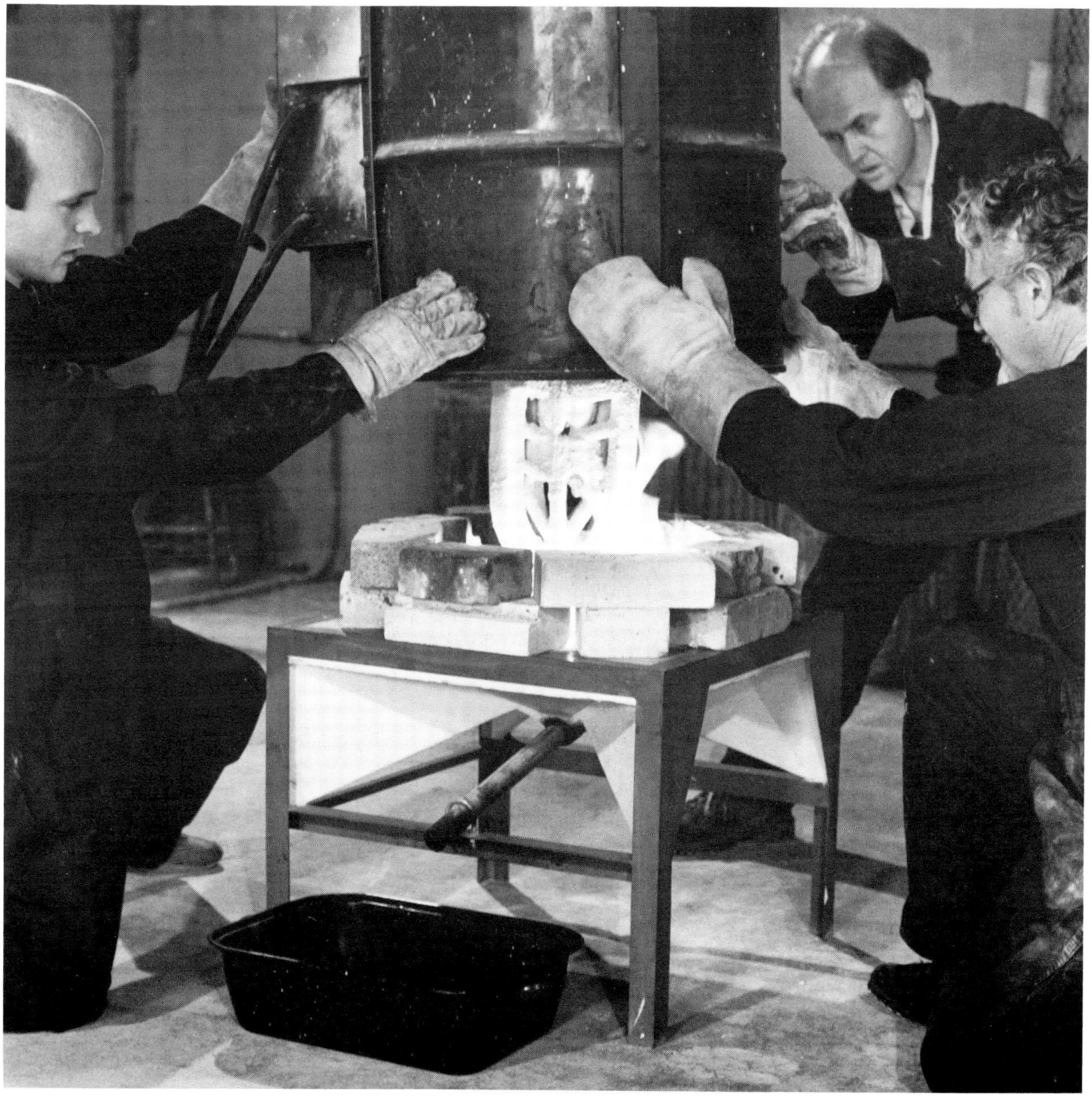

◀ 4.1 To prepare for burnout, the electric barrel kiln is set in place on the drain base and turned on. A pan is placed under the pipe drain. In forty-five minutes the kiln has reached about 1000°F. The heated kiln is then raised with the hoist, the wax-filled mold is inserted cone down, and the kiln is quickly lowered

4.2 Within a few seconds after inserting the mold, the wax in the cone ignites. In another minute or two the heat penetrates the mold wall and begins to melt the wax inside. The outer layer of wax must be liquified by this shock heating before the central mass expands or the airdried shell will crack from the internal pressure.

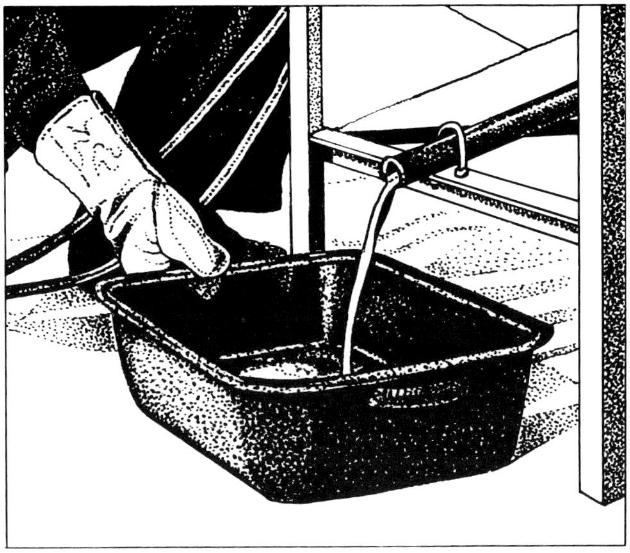

4.3 The melted wax drains out into the basin in about five minutes. Approximately ninety percent of the wax is collected to be reused. Another five minutes or so is required to burn away the wax coating the interior of the shell and the drain base. When the flame dies out, the top kiln vent is sealed and the ceramic shell mold is fired for about thirty minutes, or until the shell is free of carbon.

The most startling contrast between the traditional solid-investment approach and the ceramic shell approach to casting sculpture lies in the method of burnout, or wax elimination. Shock heating at high temperature is required for ceramic shell molds, while a long, gradual heating and cooling period is necessary for solid, plaster-based molds. Ceramic shell molds are burned out (which includes both dewaxing and firing the shell) in a preheated kiln in less than one hour, compared to the six-day cycle we use for solid-investment molds.

Furthermore, most of the wax from ceramic shell molds is reclaimed, another considerable saving, while the wax in solid molds is completely consumed in the burnout, passing invisibly into the atmosphere as a pollutant. And if the pouring has to be postponed, ceramic shell molds can be stored indefinitely for pouring at a later time. Solid molds, on the other hand, are weakened by the burnout and once cooled, are too fragile to be reheated.

4.4 At the end of the firing period, the kiln is lifted to reveal a pure white shell heated to incandescence. With the hot kiln suspended above, the glowing mold is cautiously removed from the drain base and set down nearby. Because the fused silica is now practically shock-resistant, it is able to withstand this sudden cooling which would cause ordinary ceramic ware to shatter.

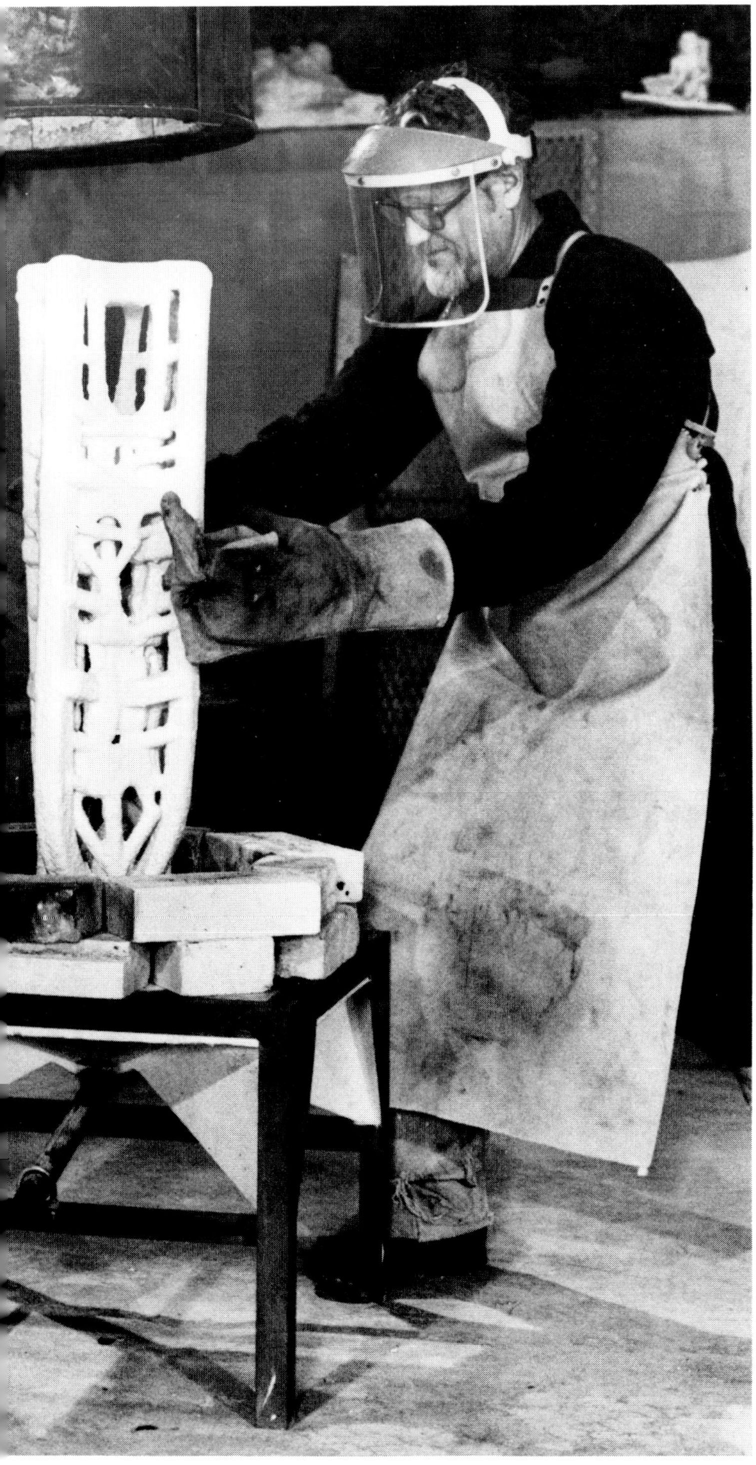

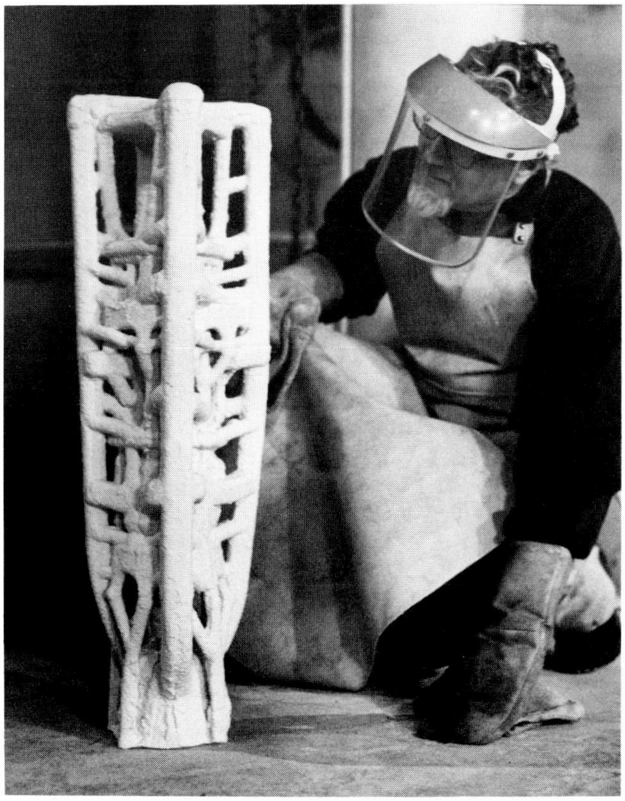

4.5 The mold is inspected very carefully for any cracks or defects. If any are found, they must be repaired before the shell can be poured. This is done with rapid-setting kiln mortar or silica-based compounds, but all repairs must be completely dry before proceeding. This mold came through the burnout without a single flaw.

5 PREHEATING THE SHELL AND POURING THE BRONZE

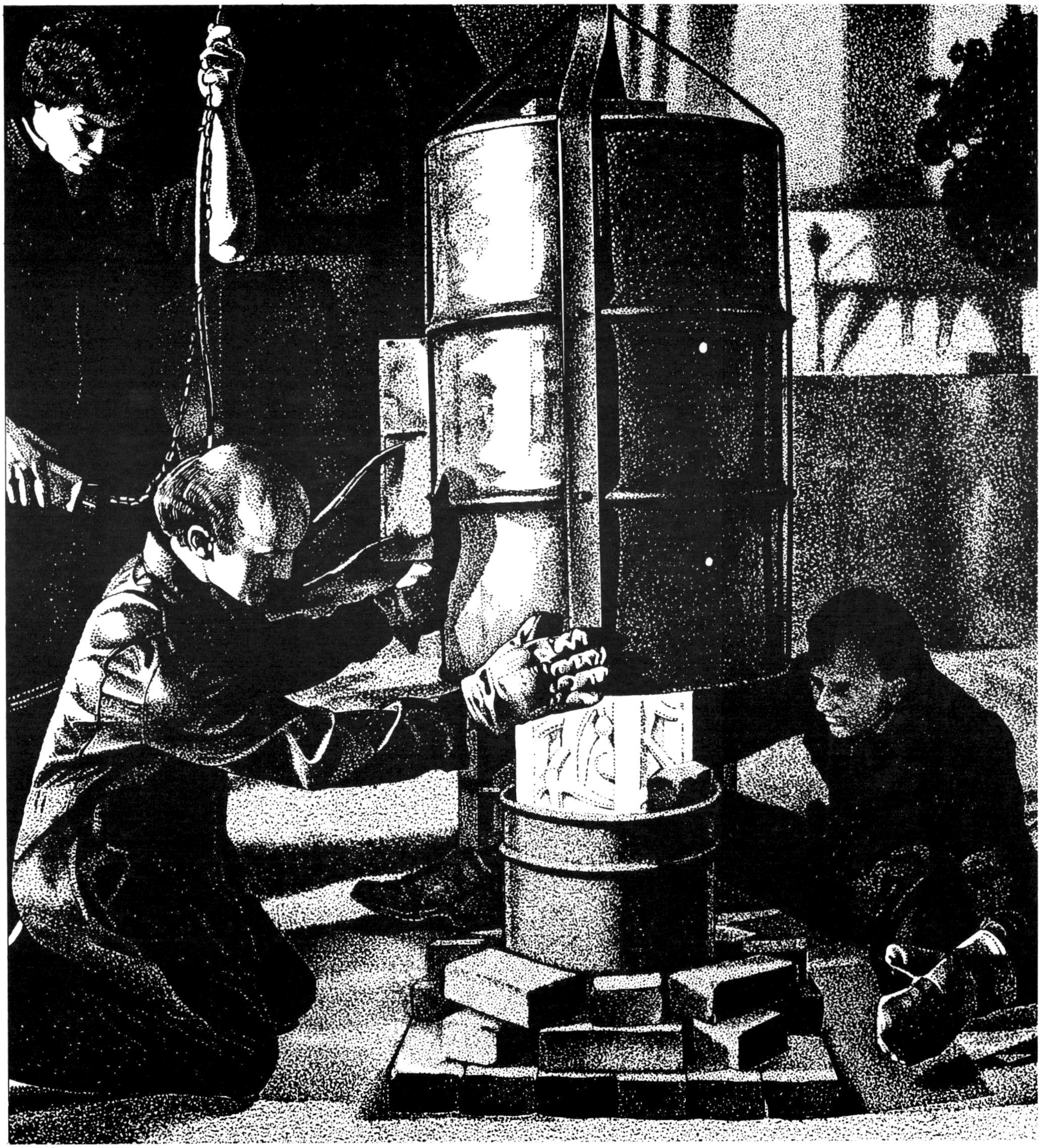

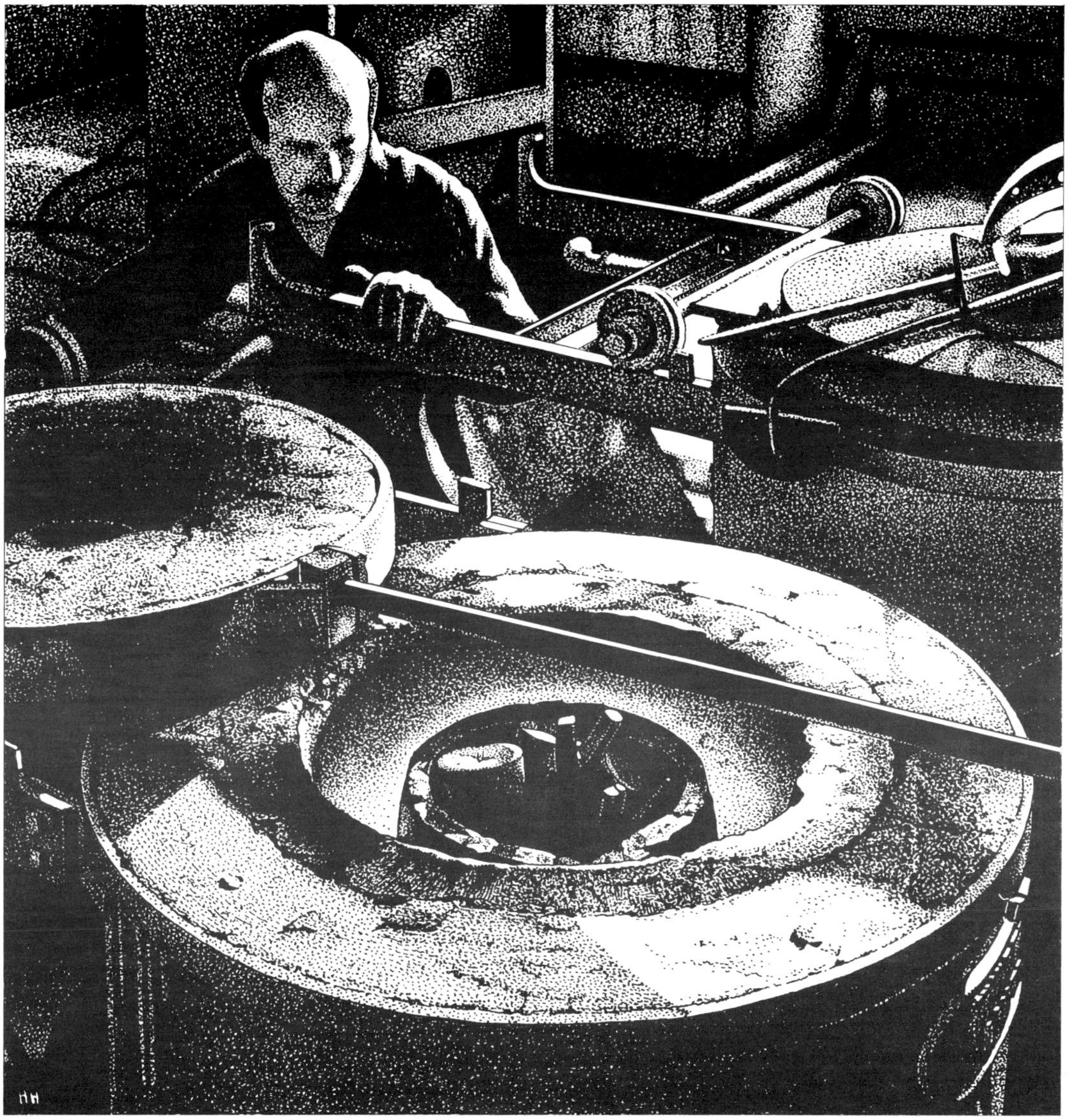

◀ 5.1 The fired shell is prepared for pouring by propping it upright in a bed of sand, braced with firebricks for stability. The same barrel kiln used for the burnout is now lowered over the ceramic shell mold and turned on. To bring the cold shell up to a red heat takes about 45 minutes, which is just about the same time it takes to melt the crucible of bronze. As soon as the preheating of the shell mold is underway, it is time to melt the metal.

5.2 The crucible is first inserted into the cold furnace, then charged with ingots and scraps of bronze. The gas is turned on and immediately ignites with a burst of flame. By opening a valve to the rear of the furnace, forced air is introduced into the furnace chamber. Here the air is being adjusted to give the proper color and sound to the flame for maximum heating. The lid will then be pulled to the right with the steel rod to cover the furnace.

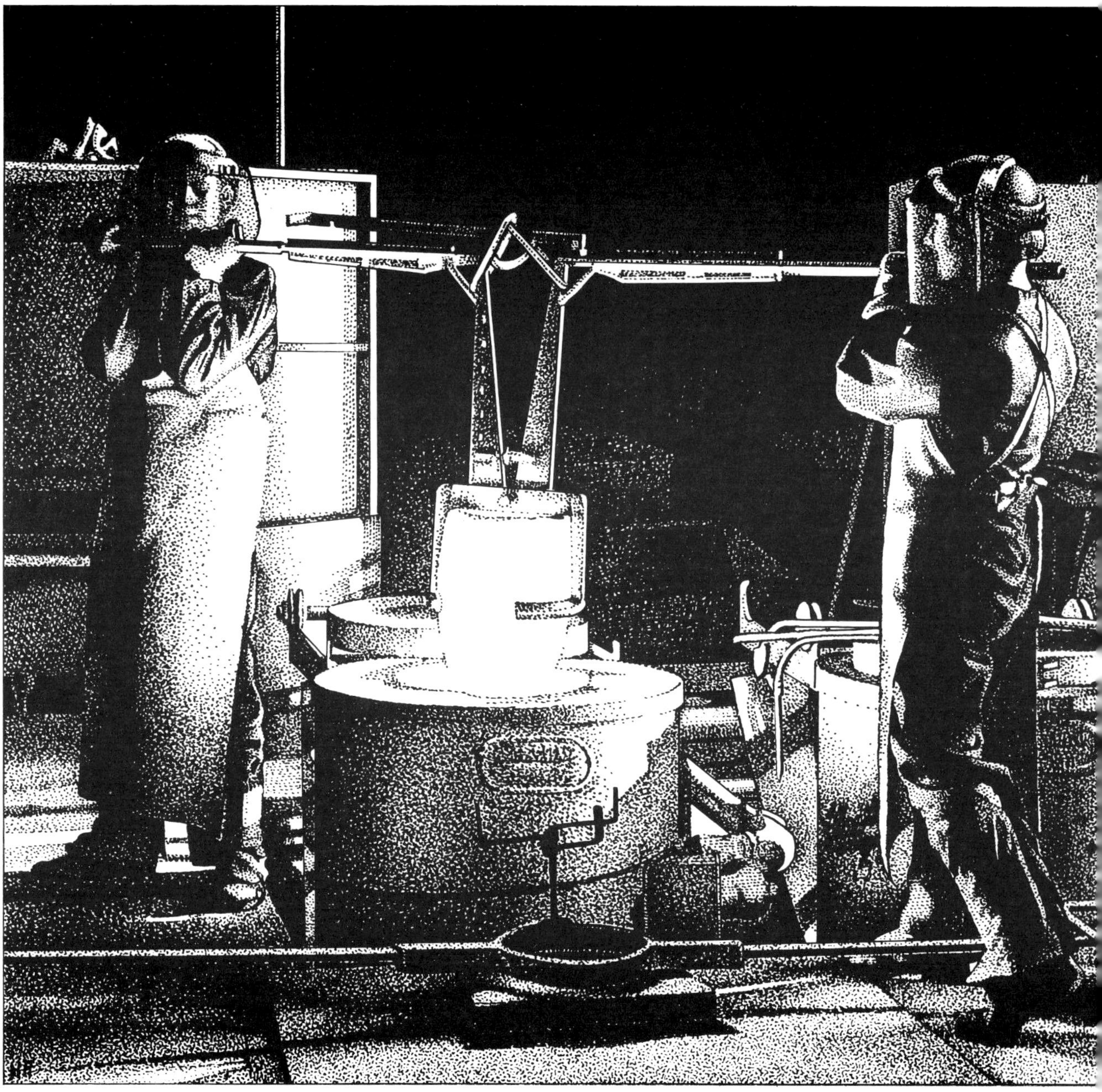

Pouring bronze is often done at night so that the founder can more accurately judge the temperature of the molten metal by its color and surface pattern. The bronze must always be heated to a higher temperature than required for pouring, to allow for the cooling that occurs while the metal is being skimmed and readied for the pour. During the brief period between removing the preheat kiln from the mold and pouring the bronze, the shell cools slightly. It retains enough heat, however, to keep the metal fluid, thus insuring that the mold will be completely filled.

Safety gear is essential to protect all foundry workers from radiant heat as well as from possible splashes or spills of molten metal. Protective clothing should include long-sleeved jackets, sturdy shoes, face shields, and leather leggings, aprons, and gloves.

◀ 5.3 It takes almost an hour to reach the desired melting temperature of the silicon bronze, which is 2100 to 2200°F. When the metal is ready, air and gas are shut off, and the tongs are lowered into the furnace chamber. The glowing crucible is very carefully lifted out, then set down inside the ring of the pouring shank.

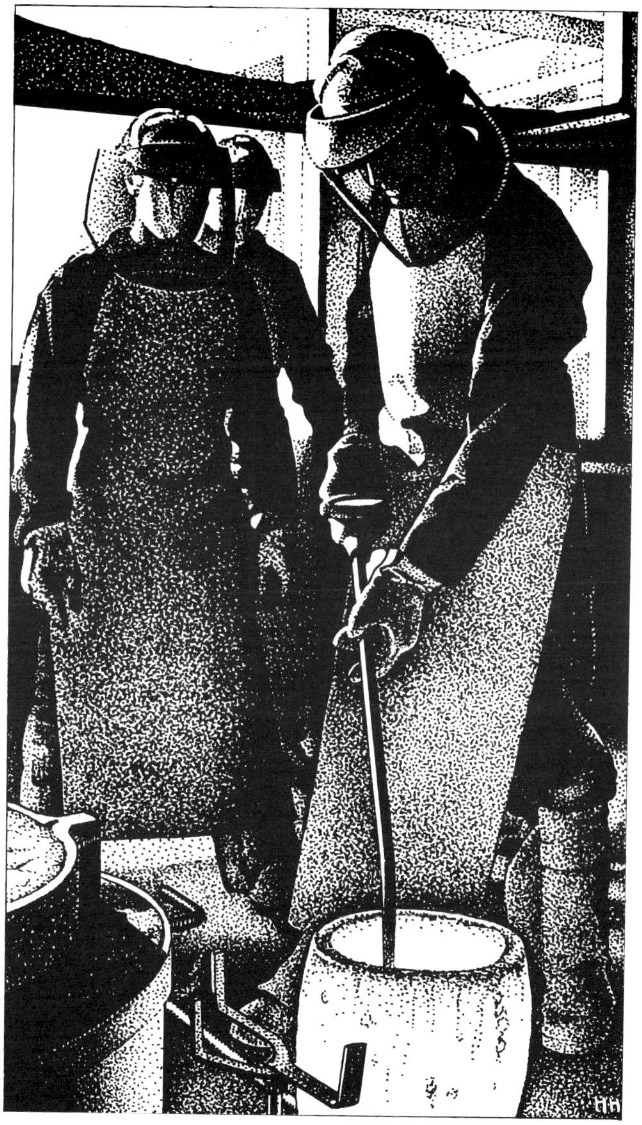

5.4 The impurities that float to the surface of the molten bronze have been skimmed off with a preheated steel skimmer. Here a pyrometer is being used to measure the exact temperature of the metal, which registers on a dial near the handle. The barrel kiln is then raised, exposing the red-hot mold for pouring.

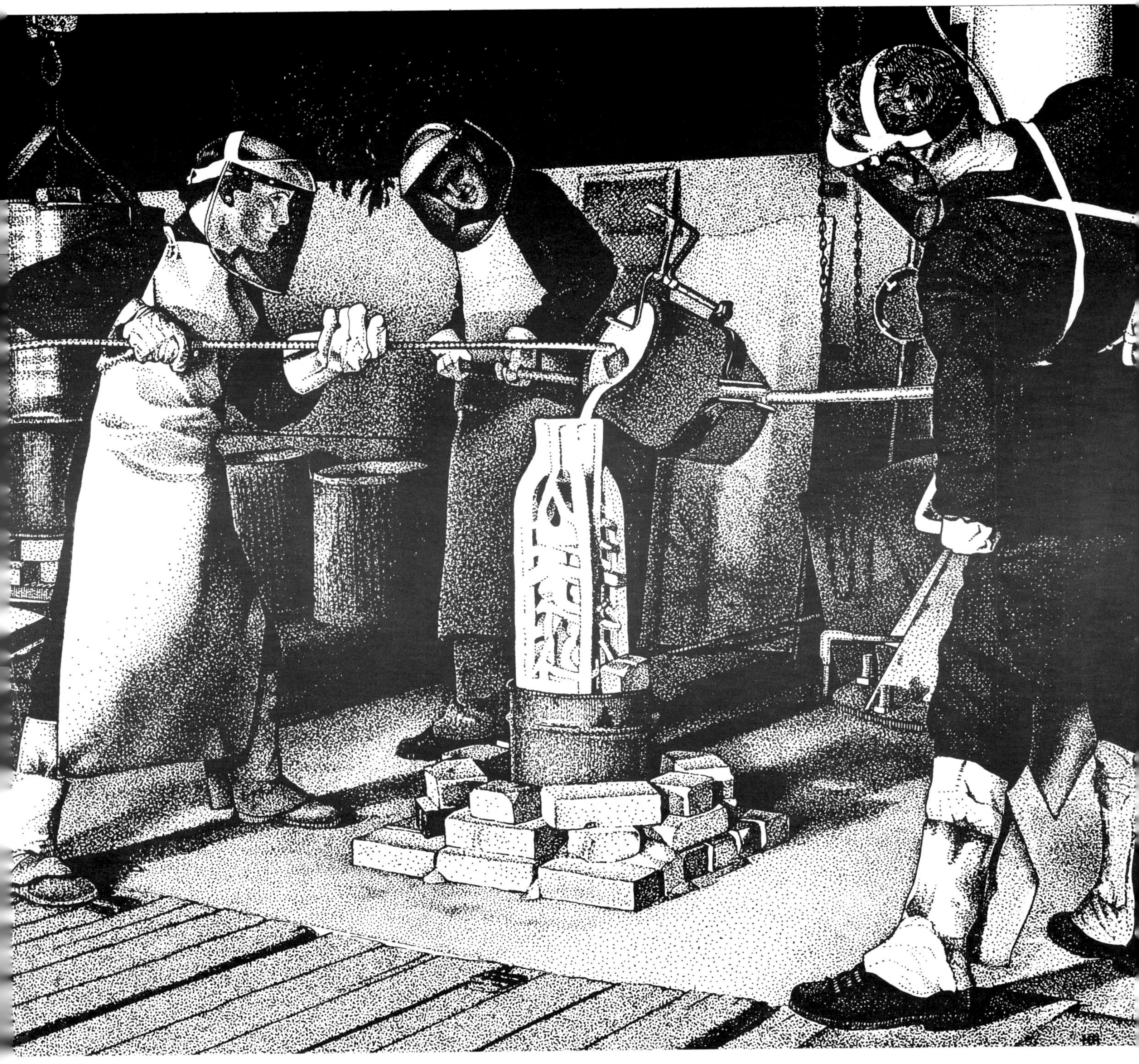

◀ 5.5 When both metal and mold are ready, the pouring shank is raised to engage the crucible, which is locked in place. The person on the single handle of the shank holds it steady while the person on the right tilts the double handles to initiate the flow of metal and control its speed. A skimmer is held in readiness by a third member of the team to make sure no traces of slag enter the mold. In less than ten seconds the shining stream of molten bronze has filled the mold.

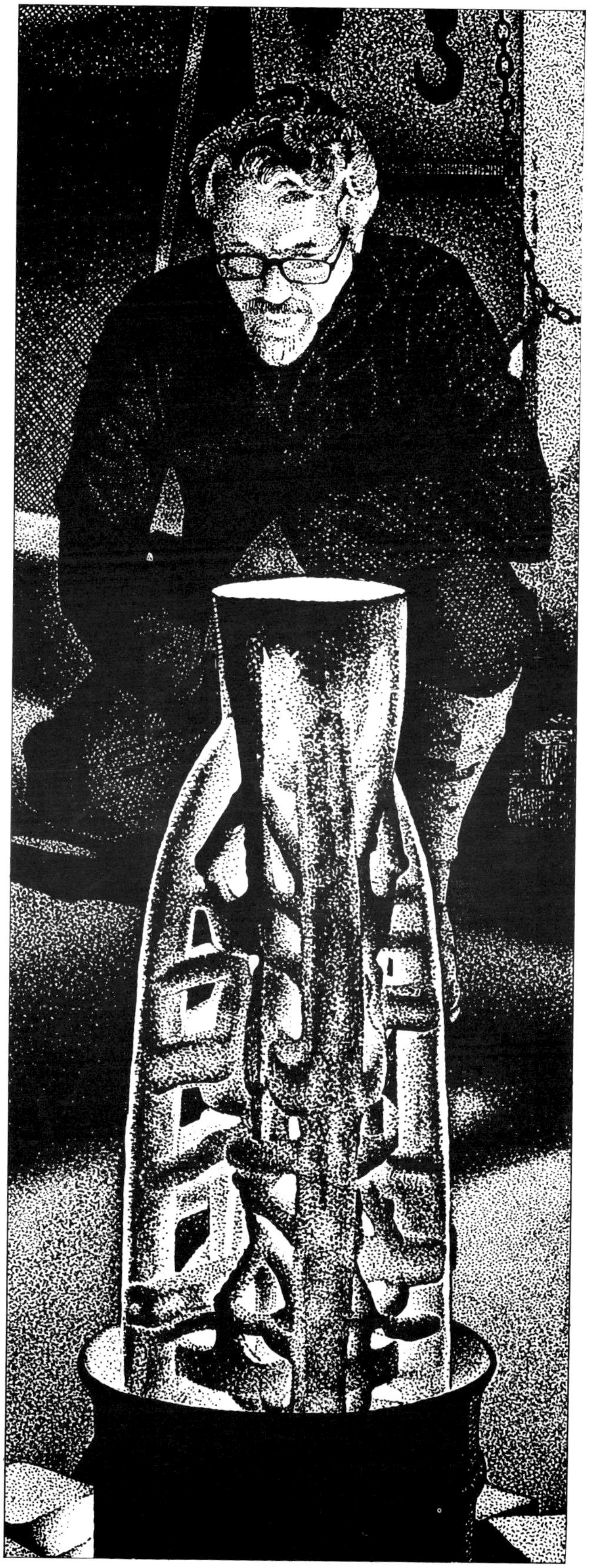

5.6 The pristine white shell, now completely filled with the fluid, pale gold metal, begins to radiate a similar golden glow. The sprue system becomes a neon network encircling the sculpture within. The heat and color, intensified in the interior spaces, are reflected from the author's face as he inspects the poured mold.

6 BREAKING AWAY THE CERAMIC SHELL MOLD: DEVESTING

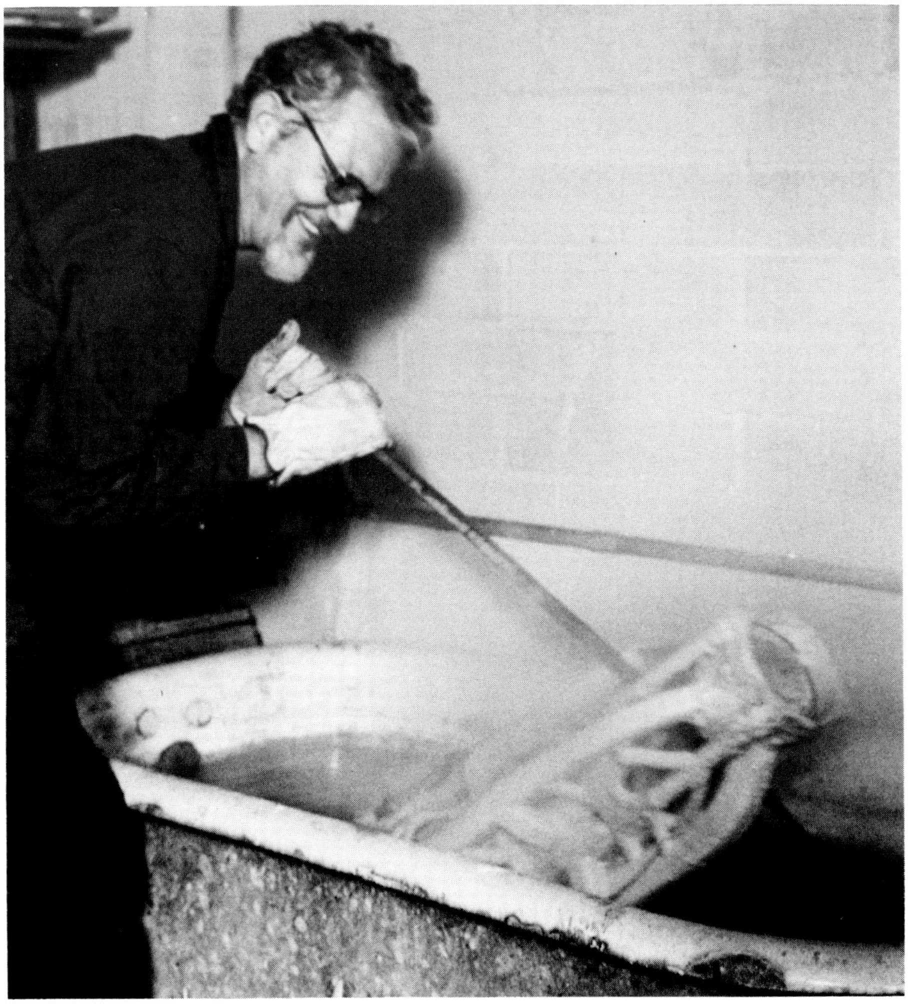

6.1 When the metal in the cone has cooled to a dark red, the bronze-filled mold is carried to the tub and slowly lowered into the water. This quenching anneals the casting, making it easier to work and less susceptible to shrinkage fractures.

6.2 As the metal cools and shrinks, large portions of the mold material fall away from the casting. Since the shell is quite brittle, gently tapping it with a small hammmer will remove most of the mold except that deep in the interior of the sculpture.

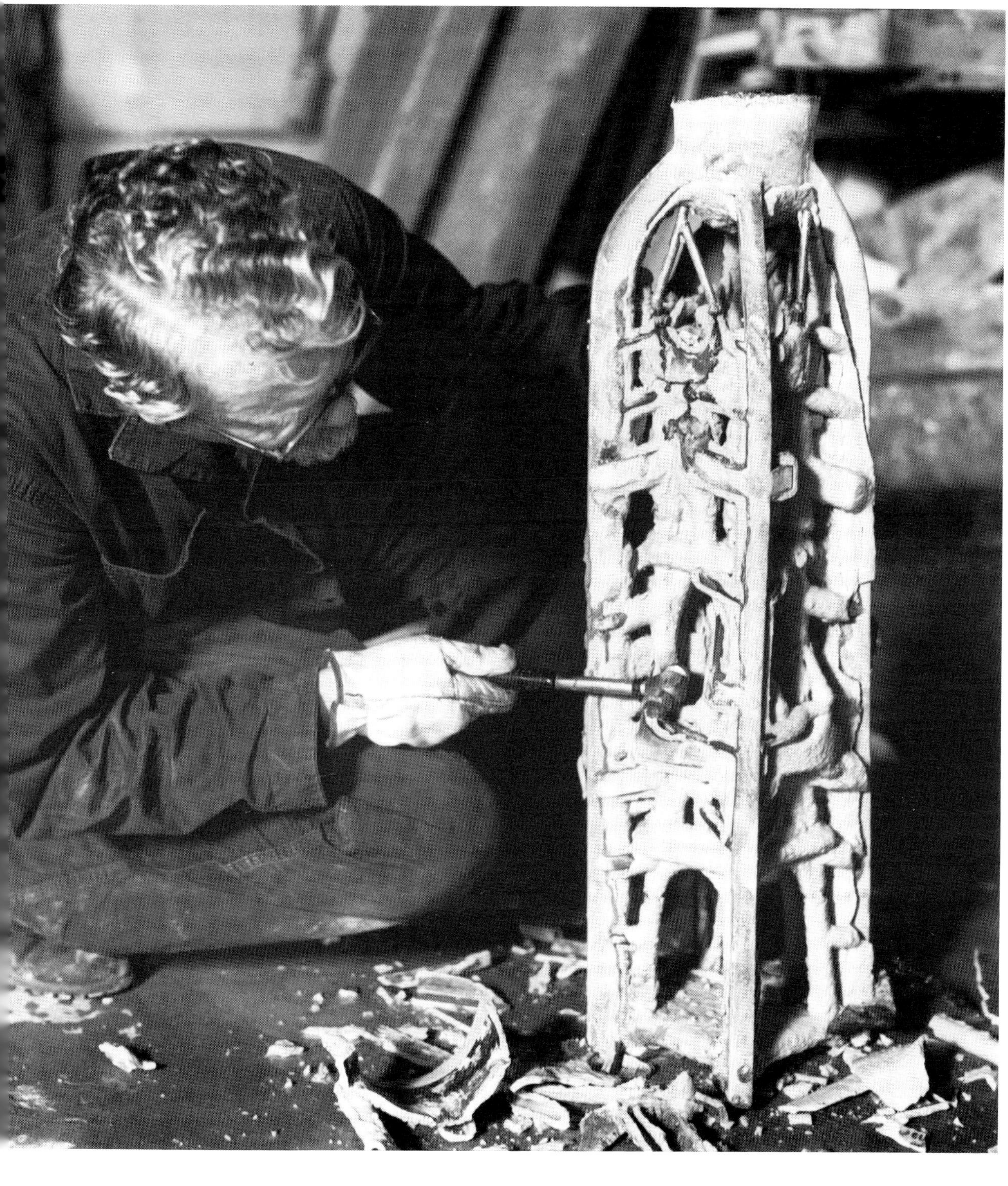

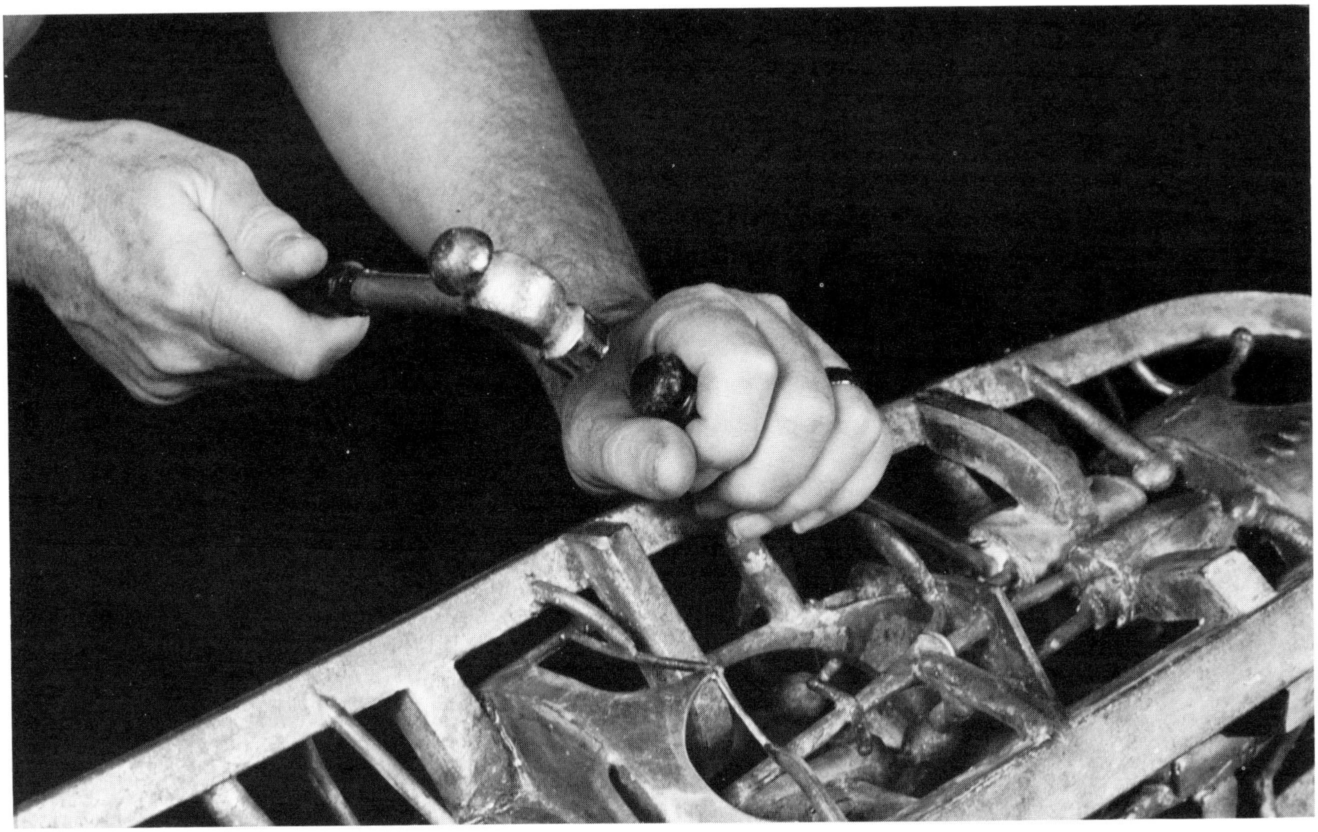

6.3 A dull screwdriver is an effective tool for reaching into mold-filled recesses and interior spaces which the hammer cannot strike. This approach is being used here to chip out the mold material filling the hollow lower half of a robed figure. For narrow spaces and incised lines, old dental tools sharpened to a long, tapered point seem to work best. Particles of the shell mold remaining on accessible surfaces may be removed with a wire wheel in an electric drill.

Besides the manual approach described here for removing the ceramic shell mold material from the metal castings, several other techniques are sometimes employed in commercial and industrial casting operations. Some foundries subject the mold-covered casting to intense vibration, using power-driven equipment similar to a jackhammer to crack and loosen the brittle shell. Another approach used in some mass-production lines is to bombard the mold with a jet of water under extremely high pressure. Because of the potential danger to the operator, this is usually done in an enclosed booth. Still another method utilized in industry is to dissolve the shell mold by immersing the casting in a chemical solution that will not affect the metal.

These industrial processes, however, all require costly, heavy-duty equipment and elaborate safety precautions for dealing with dangerous machinery or toxic chemicals. Such processes are neither appropriate nor economically feasible for a campus sculpture foundry or an individual sculptor's studio. For these situations, the one technique that can be recommended to supplement the manual approach is blasting with glass beads, which should be done only after most of the mold has been removed manually. Beadblasting will be discussed in a later chapter.

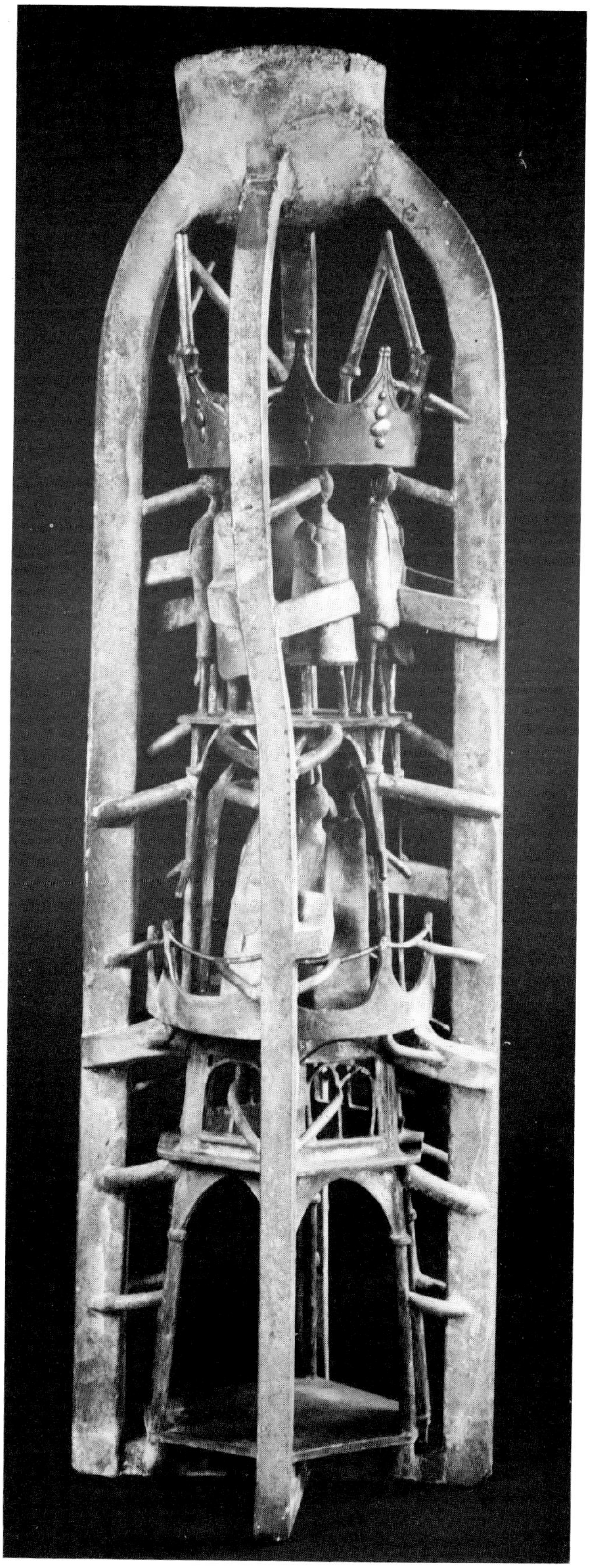

6.4 The bronze casting with all of the ceramic shell mold removed is an almost exact duplicate of the sprued wax. The only variation is in the vents; all of them did not fill in all the way. This does not matter, however, since the sculpture cavity itself is completely filled with metal.

7 REMOVING SPRUES AND CHASING THE CASTING

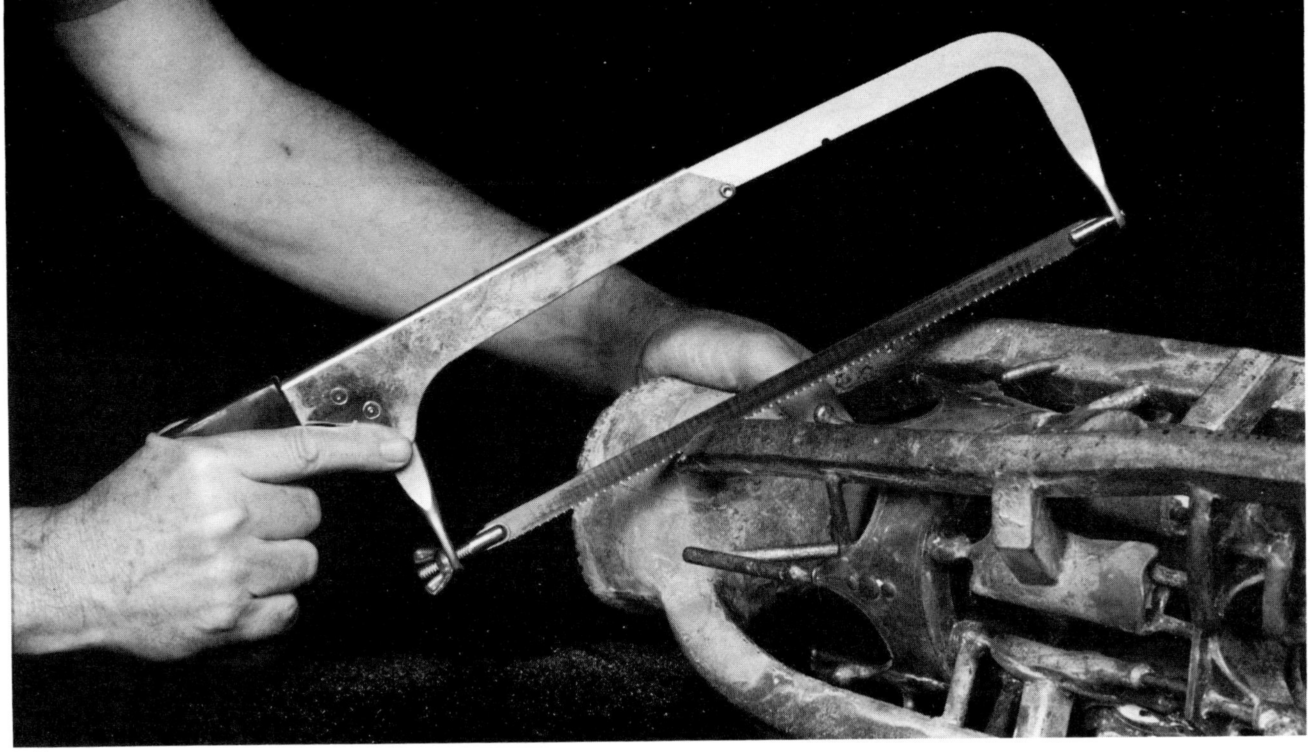

7.1 The wax rods that formed the sprueing and venting systems for the mold have now been transformed into bronze, and must be removed. The simplest approach is to saw the sprues off with a hacksaw, but this is also the most laborious way to remove them. If any trace of mold material remains in the path of the saw, even the toughest hacksaw blade will be immediately dulled by contact with the ceramic shell. The bronze casting should be securely fastened in a bench vise to prevent movement during the saw stroke.

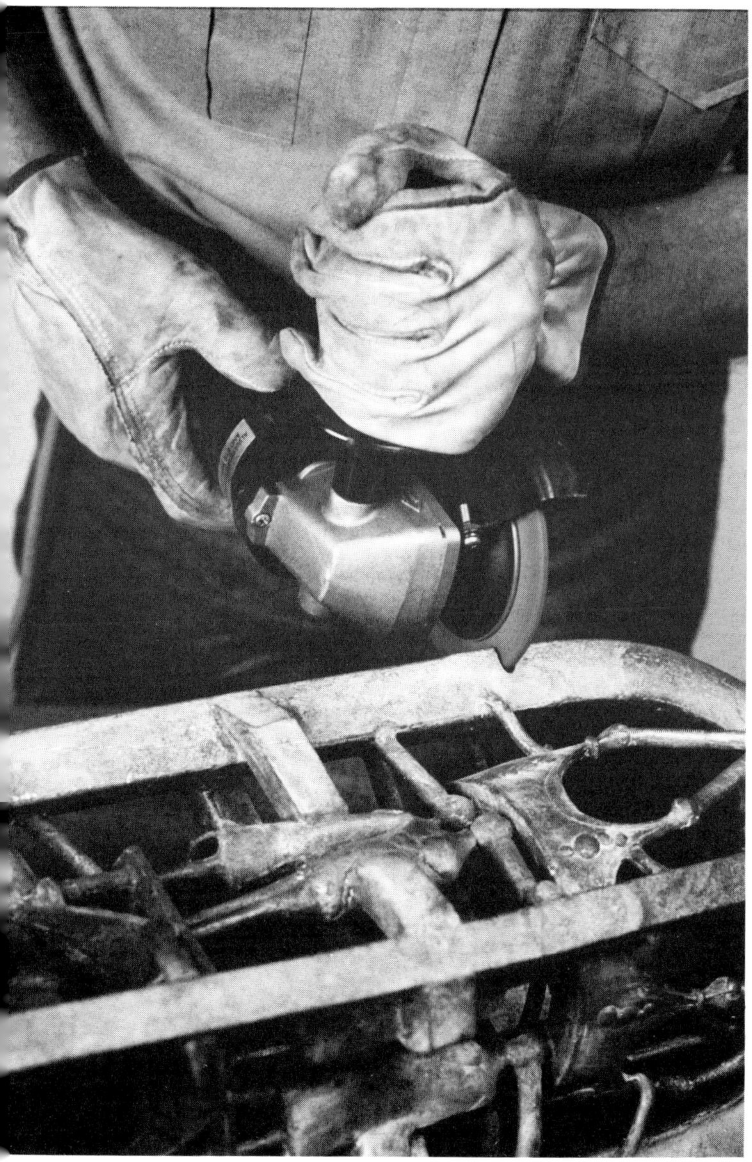

7.2 The fastest way to remove the bronze sprues is to use a portable grinder with an abrasive cut-off wheel. The tool must be tightly gripped with both hands as the wheel enters the metal. Safety goggles (or a face shield) are essential to guard against flying particles of metal; ear protectors are needed to cope with the high noise level.

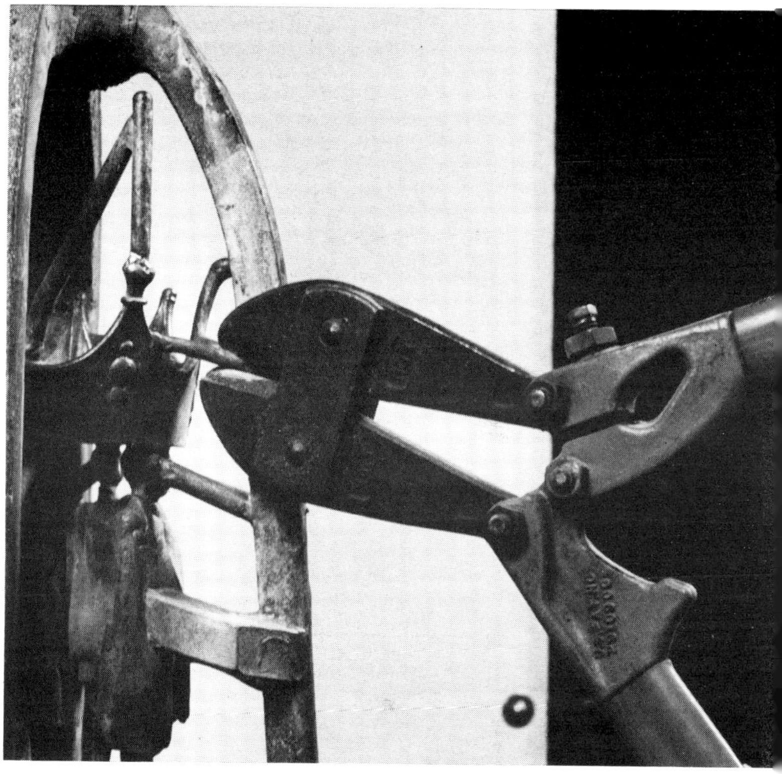

7.3 Vents and small sprues ¼″ to ⅜″ in diameter can be cut off with a bolt cutter. Under pressure from the dull, leveraged jaws, these thinner metal rods snap apart. Because of the force exerted sideways from the cut, however, one should avoid using this tool to cut sprues attached to fragile sections of the sculpture.

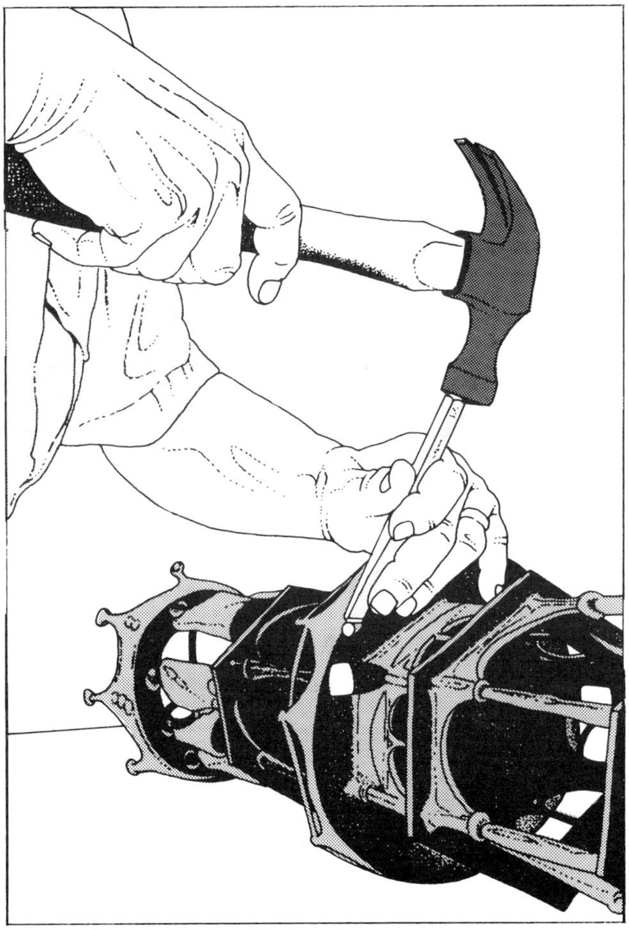

7.4 After the sculpture has been freed of the sprue system surrounding it, the stub of each separate sprue must be removed cleanly from the sculpture's surface. In this process, called chasing, all evidence of that connection must be obliterated. On sections of the sculpture strong enough to withstand the impact, these stubs of bronze may be chiseled off close to the surface with a cold chisel.

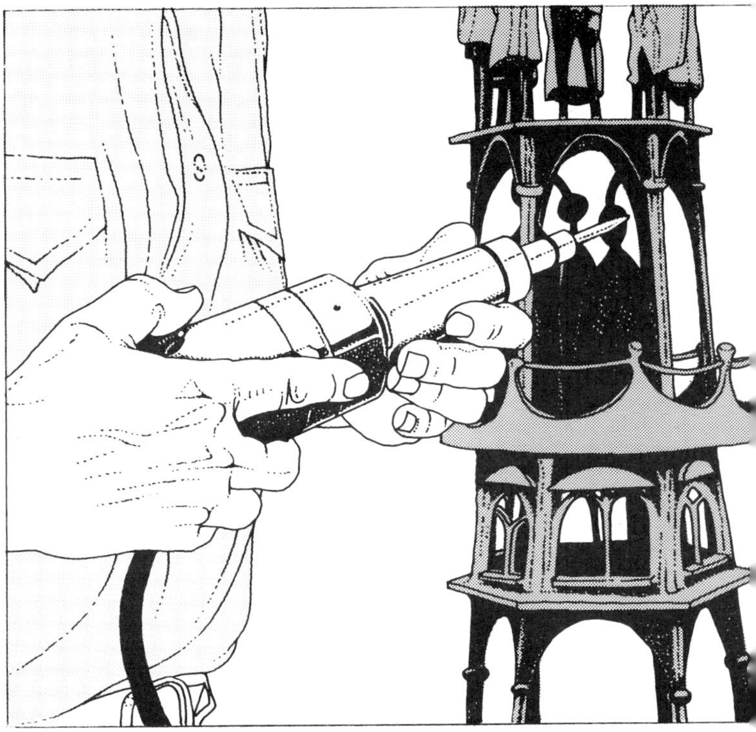

7.5 Sprues that have been sawed or chiseled off must now be filed down as close as possible to the surface of the sculpture. This may be done with hand files or, as shown here, with a rotary file in a high-speed electric die-grinder. In either case, work progresses from coarse to finer files until no trace of the sprue connection remains.

7.6 The final step in chasing is to make the bright, smoothly filed spots where the sprues were removed blend with the texture of the surrounding surfaces. Here an adjustable-speed, flexible shaft tool is being used to texture one such area. At a low speed controlled by the foot pedal, a worn rotary file is stroked in various directions, as one does in cross-hatching, until the smooth spot merges imperceptibly with the surface on which it lies.

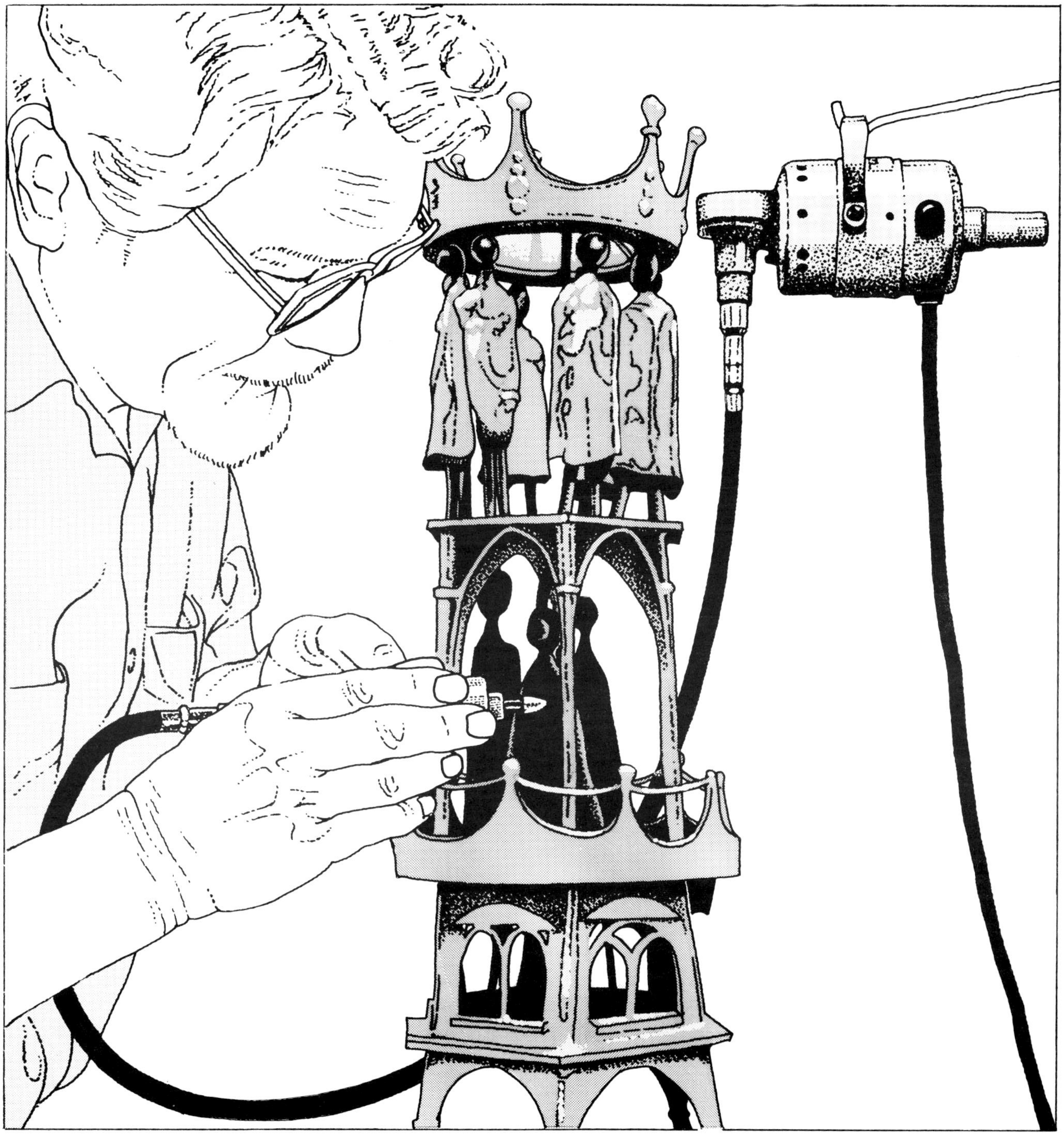

8 CLEANING AND COLORING THE BRONZE: PATINATION

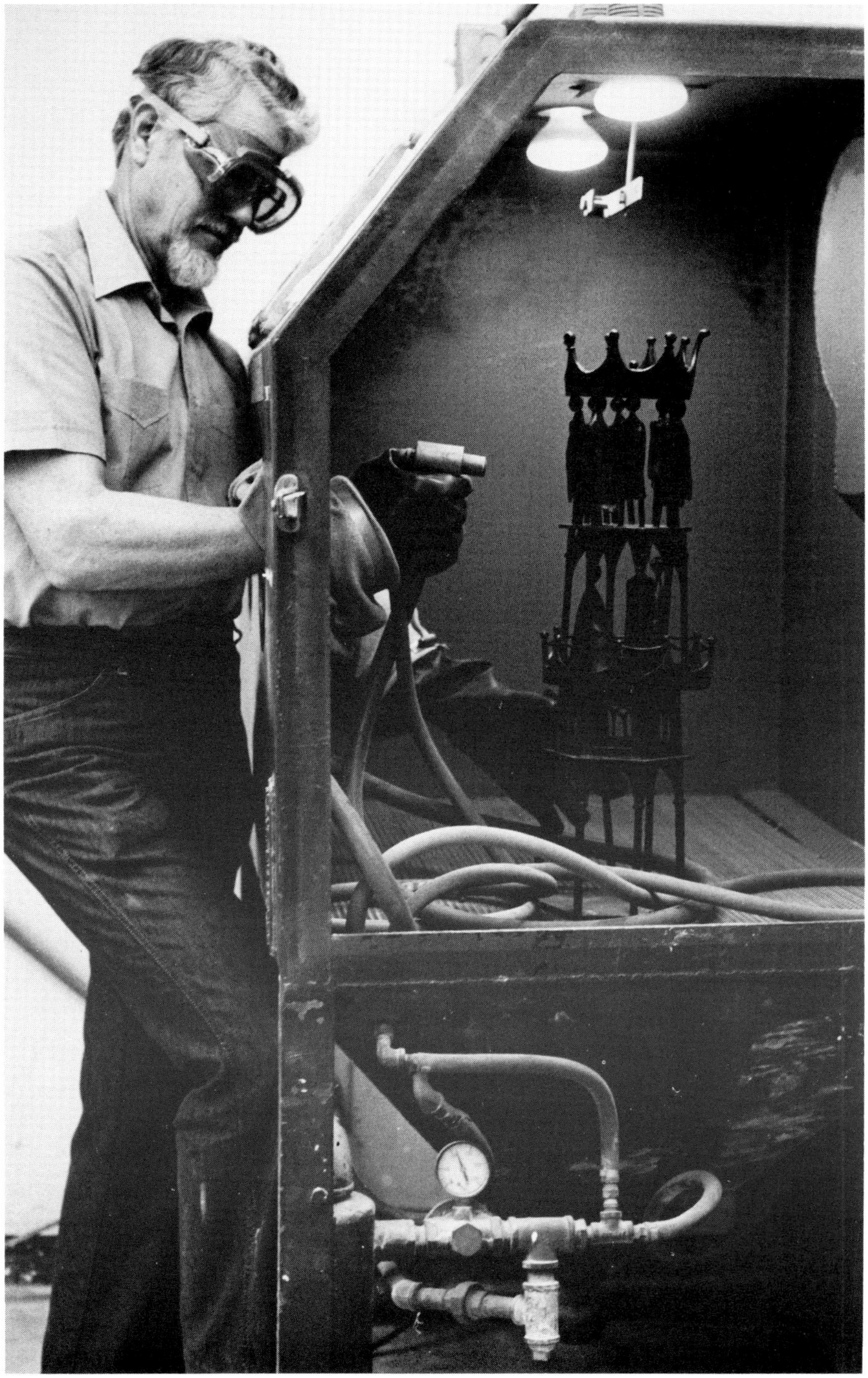

The application of a patina to a bronze sculpture accelerates artificially the gradual coloration process that would occur naturally if the work were exposed to the elements over a long period of time. Different chemical solutions can be applied to the metal in a variety of ways to produce a range of colors. The solutions are usually brushed or sprayed onto a heated bronze. However, the sculpture can also be immersed in a hot or cold chemical bath, exposed to chemical fumes in a sealed atmosphere, or buried in sand or sawdust saturated with chemicals. The ultimate intent is to enhance the sculptured form and emphasize its surface qualities by adding tonal and hue variations to the otherwise even-colored metal.

In order to control the patina that is to be applied, the bronze must first be thoroughly cleaned by either a chemical or a mechanical means. The traditional way has been to immerse the casting in a dilute acid bath. However, any trace of acid remaining in a crevice or porous area can continue to attack and discolor the bronze. After experiencing this result, I abandoned the chemical approach in favor of mechanically cleaning my bronzes. Since sandblasting erodes the surface, I now use in the sandblaster only fine glass beads, which clean the metal without abrasion. After cleaning, the sculpture should be handled with soft, cotton gloves to prevent oil from the skin from affecting the application of the patina.

Although there are many patina "recipes" combining lots of different chemicals, some groups of chemi-

◀ 8.1 Before the patina can be applied evenly, all traces of mold material and oxidation must be removed from the surface of the casting. The best way to clean the bronze is to blast it with fine glass beads, which do not abrade the surface as sand would do. The sculpture is in place in the cabinet, ready for the door to be closed. When the foot pedal is pressed, the glass beads will be shot from the nozzle.

8.2 After the sculpture has been heated to about 300°F. in the barrel kiln, the bronze is sprayed with a fine mist of cupric nitrate solution. Four to six applications may be necessary to get the color desired. The spraying should be done outdoors or in a well-vented spray booth; a face mask with filter is advised.

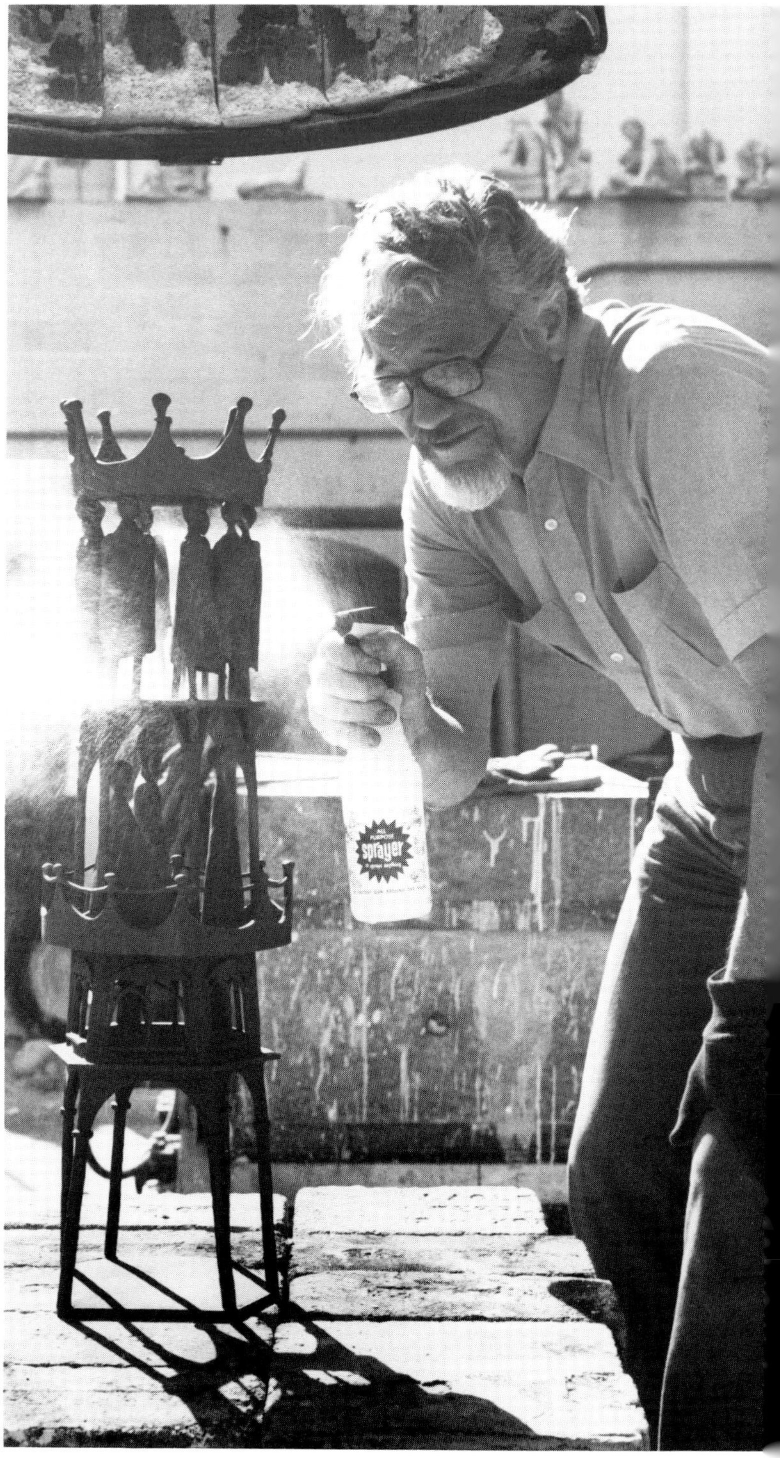

8.3 When the spray applications have produced a satisfactory color on the bronze, the coating that has built up on the surface usually needs highlighting with steel (or bronze) wool. This slight thinning of the patina in various areas will restore faint glimpses of the warm bronze underneath to emphasize the form, bring out textures, and eliminate a flat, painted look.

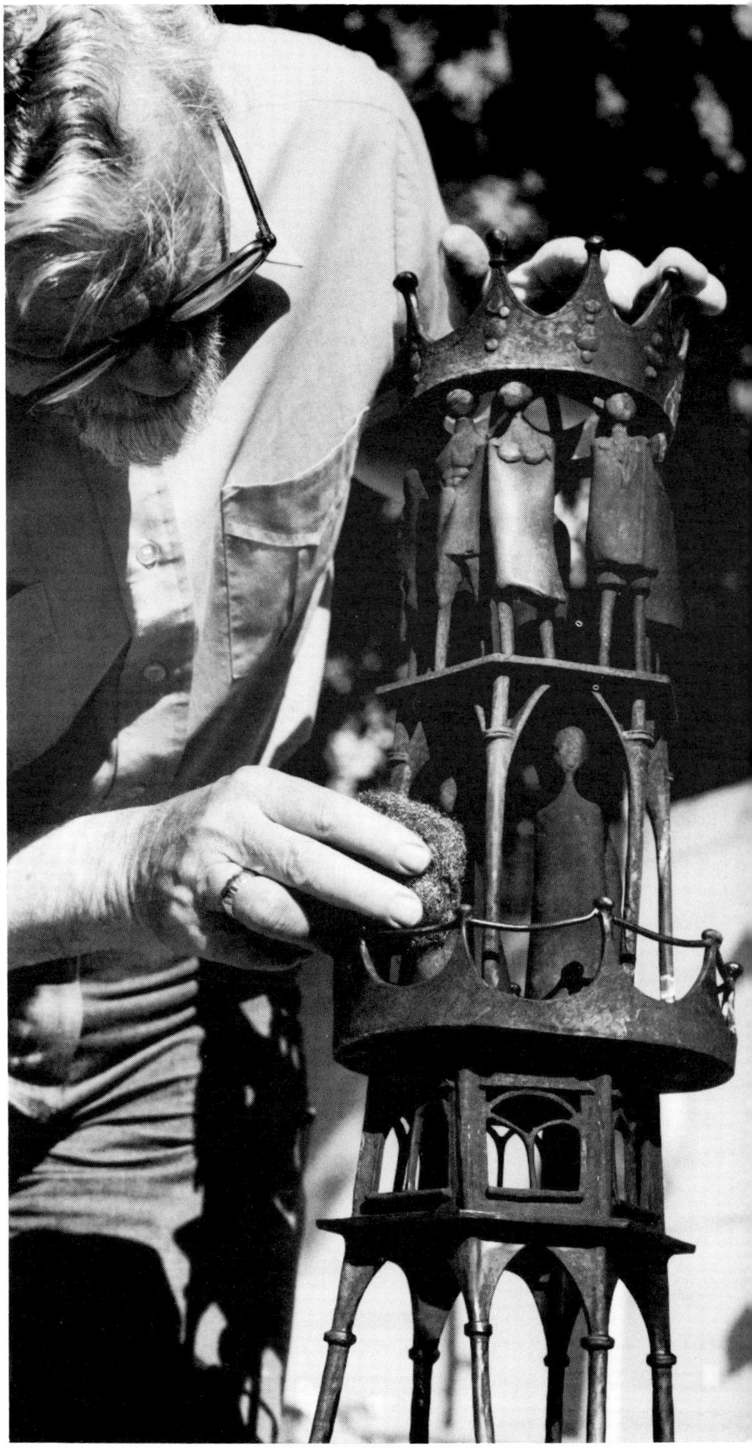

cals should be avoided. Some produce patinas that are powdery and thus too transient; others are not safe for the metal. I avoid all chlorides, for example, which can combine with atmospheric moisture to form an acid which will corrode the bronze. Some chemicals should be avoided because of their toxicity—compounds of cyanide and mercury, for instance. To simplify the patination process and minimize both the hazard and the cost, I prefer to use only four chemical compounds, all available in crystalline form: liver of sulphur, ferric nitrate, cupric nitrate, and sodium thiosulfate.

To get an even patina, the sculpture must be preheated evenly to about 250° to 350°F. before a brush or spray application. Large works can only be warmed with a torch, but smaller pieces can be heated more evenly in a kiln or kitchen oven. The metal should be just hot enough for the chemical solution to sizzle and evaporate instantly. When the solution begins to run, stop spraying and reheat the bronze.

These four chemicals will provide a reasonable spectrum of patina colors ranging from black and brown to greens and blues. Solutions of the first three compounds should be brushed or sprayed on. *Liver of sulphur* will produce a cold black. (Dissolve a yellow lump about half the size of a walnut in a pint of water.) *Ferric nitrate* will give a warm, rusty brown tone (one teaspoon of the pale lavender crystals to a pint of water). *Cupric nitrate* alone will produce a turquoise blue (one walnut-sized blue lump to a pint of water). By cautiously spraying ferric nitrate onto a base (several coats) of cupric nitrate, however, an entire cool-to-warm spectrum can be achieved. The blue can change to an apple green, then to olive or silvery grey as the two compounds are blended by alternating spray coats (reheat the bronze before each application).

For pieces small enough to fit into an enameled pot, *sodium thiosulfate* (hypo—prismatic type, pentahydrated—from Kodak, for developing film) can give a warm, golden to chocolate brown patina with a purplish iridescence that regrettably fades away. About half a cup of the clear crystals in a gallon of water will make a solution into which the sculpture can be immersed. Boil gently for half an hour or more. The bronze will darken gradually, and should be removed and rinsed when the desired color is reached.

After highlighting with steel wool, as shown in Figure 8.3, the entire sculpture must be sealed to prevent any further oxidation or color change. The bronze must first be absolutely dry. It is then coated with either a thinned lacquer or acrylic spray, or more traditionally, a hard wax brushed on the warmed sculpture.

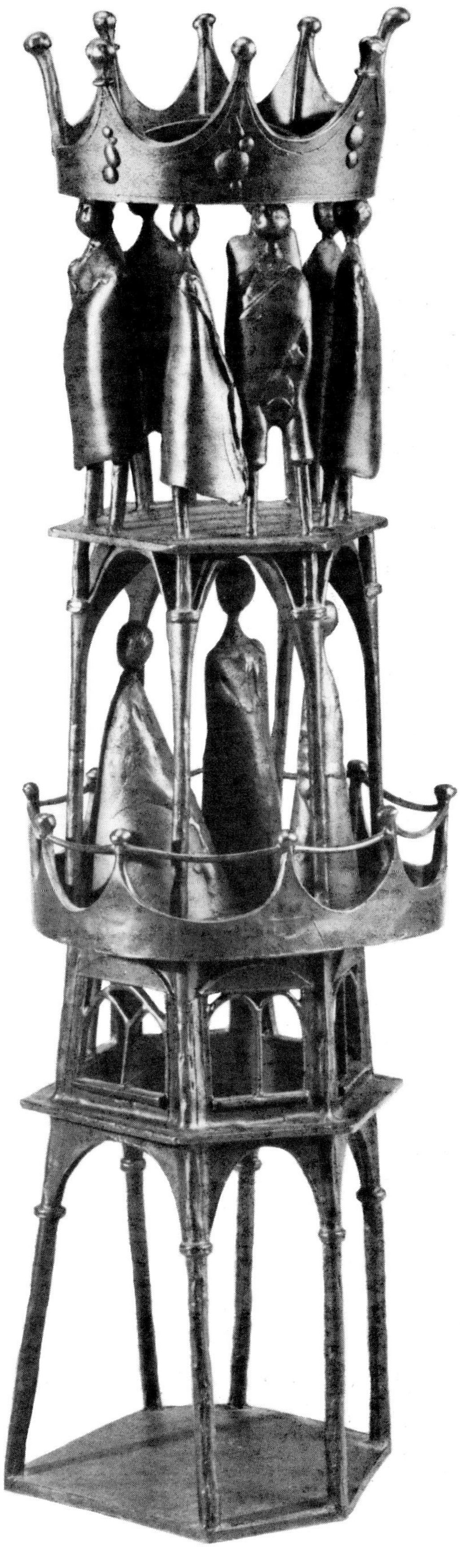

8.4 The finished bronze, *God Save The Prince II,* is just under 25″ tall and weighs 23 pounds. It has a warm black patina preserved by a thin spray coating of semi-matte lacquer. The ceramic shell mold reproduces with such high fidelity that the bronze exactly duplicates the original wax surface, differentiating between areas that were modeled, cut or burnished with tools, and fused with a torch.

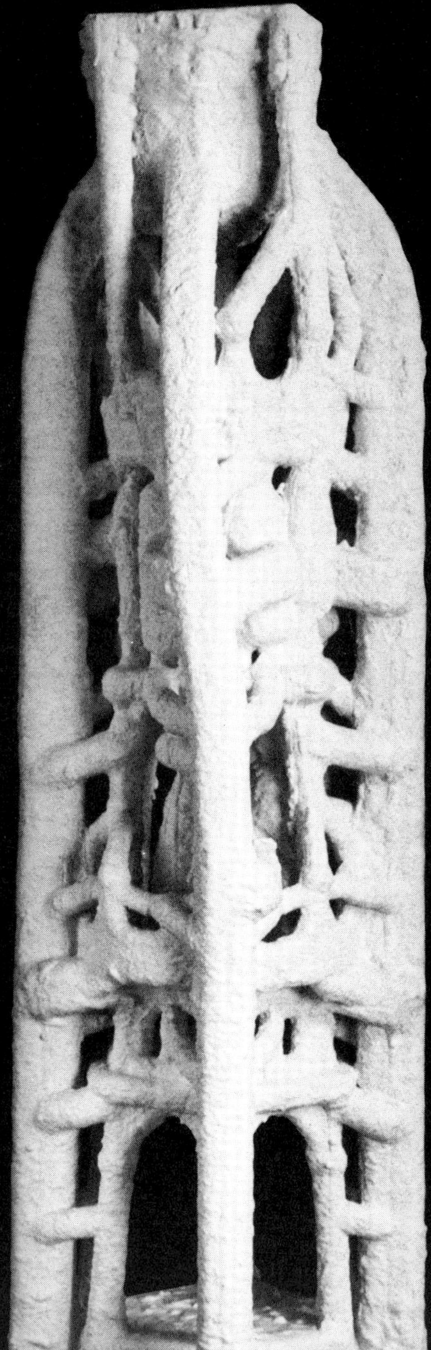

PART TWO

CERAMIC SHELL MOLDMAKING FOR THE STUDIO

A Simplified Approach

9 PREPARING THE STUDIO FOR CERAMIC SHELL MOLDMAKING

In the first section of this book I have attempted to define in visual terms the various stages in casting a bronze sculpture from ceramic shell molds so that anyone can understand the process. This second section has been written expressly for those who may wish to utilize the process themselves or introduce it to their students. If we had the luxury of setting up a complete new foundry, each of the stages could be planned for a separate room. My concern here, however, is for the college or university program where perhaps only one studio/classroom is available to students and instructor for the entire range of sculpture processes. Of equal concern is the individual sculptor whose single studio space must encompass as many steps in the process as possible. We need to consider at this point which of the steps described in Part One can be accommodated in a typical studio or classroom, what basic equipment and utilities they require, and how to provide for those steps that cannot be handled within the studio or classroom.

The steps or stages in the ceramic shell process can be grouped into two major categories: those that take place before the melting and pouring (founding) of the metal, the *prefounding* stages; and those required after the metal has been poured, the *postfounding* stages. For maximum efficiency, a commercial foundry might align these two categories of work activity on opposite sides of the casting area so that a piece begins at one end of a straight production line and emerges as a finished bronze at the other. In the classroom or sculptor's studio, however, the work area for each stage of the process may depend primarily on where the existing utilities are located in the room and which materials (such as plaster) need to be in a limited area. The separate work areas required in the studio will correspond roughly to the stages outlined in Part One, but will also be defined by the material being used, such as wax or metal.

To provide for constructing and sprueing the wax sculpture, as described in Chapters 1 and 2, we must obviously have a *wax working* area, which in a foundry may also be referred to as the *patternmaking* area. Minimum requirements here are a stove or hotplate (either gas or electric) for melting pans of wax, and work surfaces (tables or counters) for pouring and handling waxes. For small tools and appliances such as soldering irons, melting pots, etc., electrical outlets are desirable but not essential, since alcohol lamps could be used. There should be sufficient floor space to allow for pouring wax in and out of molds, and a floor surface from which wax spills can be scraped.

Since plaster or flexible molds are likely to be used at some point for casting waxes, a *plaster-working* or *molding* area is needed (not to be confused with the investing area where the ceramic shell molds will be made). A sink and adjacent work surfaces must be available. A large container (plastic garbage can) of water should always be provided, of course, for rinsing fresh plaster from hands and tools. Storage containers for dry plaster and cupboards for flexible molding compounds are also required. Electrical outlets should be available so that portable hand mixers can be used if needed. The plaster area should be confined to minimize the mess and reduce the possibility of contaminating other materials.

The next essential area is the *investing* area, where the waxes will be coated with slurry and stucco to build the ceramic shell molds. Sufficient space must be allowed for (a) the slurry mixer with two containers for slurry and colloidal silica; (b) at least two boxes for coarse and fine stucco; and (c) a rack or table for drying molds. The specific equipment needed for this area will be discussed in detail in the next chapter. Although I have never been able to do so, this stage of the process should be handled in a separate room where temperature and humidity can be controlled, since hairline cracks may appear in the molds if the temperature fluctuates more than five degrees. Ideally, the drying room should remain at a constant 75°F. with a relative humidity of 50 percent.

When the ceramic shell molds are complete, they are then burned out and poured, as described in Chapters 4 and 5. These two operations must be done out-of-doors or in a well-ventilated area, and would be unsafe in most individual or classroom studios, although an outside patio adjacent to the studio could provide an appropriate site. Special equipment for the *burnout* (wax elimination and firing of the shell) will be described in Chapter 11. I will not attempt to define the equipment need for the actual *founding* stage, the melting and pouring of the metal as described in Chapter 5, since it should not be done in the studio. Anyone interested in setting up a melting furnace should either contact a manufacturer or retailer of industrial foundry equipment (see List of Suppliers) or refer to some of the books that already provide instruction on building your own furnace (see Bibliography).

The first postfounding stage is *devesting,* removing the shell mold from the casting as described in Chapter 6. This operation could be done in the outdoor foundry area where the metal was poured. If done in the studio, a special work area is not needed—only a workbench with a sturdy vise. The only caution that must be observed is to watch for pieces of the shell flying off into the surrounding area.

The later stages in finishing the sculpture involve *metal-working* processes. The first is *chasing,* described in Chapter 7, which includes removing the sprues and their connections to the work. I use the same workbench for this as for devesting, so a special studio area is not necessary. In a studio/classroom, however, the noise could be a problem for everyone else in the room, even if the person working on the metal wore ear protectors. This problem could be solved if the chasing area were partitioned off and insulated for soundproofing. If cast metal sections have to be *joined* or *repaired,* then welding equipment, a worktable, and a welding booth or screen will be needed.

The final stage is *patination.* As stated in Chapter 8, this step also should be done outside if possible. If it has to be done inside the studio, a separate spray booth should be installed so that all chemical mists and fumes can be completely vented.

In summary, the studio/classroom or the individual studio needs to have, as a minimum, separate areas for: (1) wax working; (2) plaster and flexible molding; (3) investing with slurry and stucco; and (4) metal working. Desirable improvements beyond the minimum include partitioning off separate areas for investing, with temperature and humidity control; for chasing, with soundproofing; and for patination, with complete ventilation.

10 PROCEDURES AND EQUIPMENT FOR SLURRY AND STUCCO

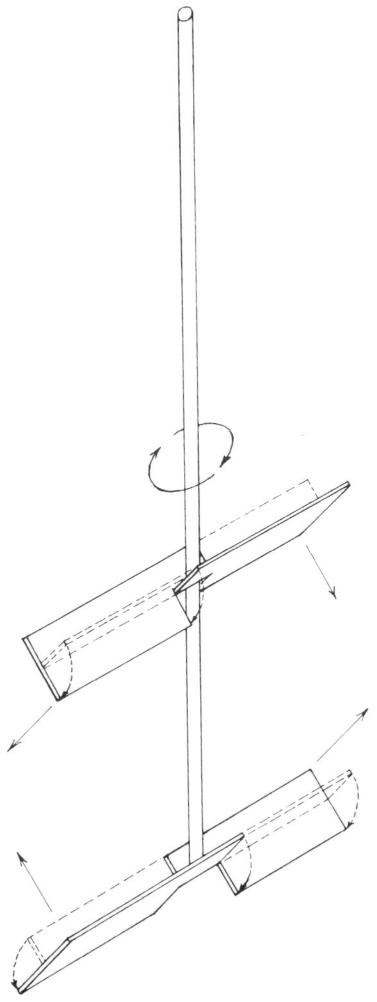

10.1 Slurry mixer made from angle iron and steel rod. Bottom blades push up, upper blades push down.

The slurry consists of very fine particles of fused silica suspended in the liquid colloidal silica sol. Unless the slurry is stirred at frequent intervals, these particles will settle to the bottom of the container and pack into a solid mass. One must choose, therefore, to either (a) mix up a new batch of slurry for every mold coat, discarding the unused portion each time, or (b) set up some device for stirring the slurry constantly, or at least often enough to keep the fused silica in suspension. The first option I rejected because it would be too tedious and time-consuming to be constantly mixing up new batches of slurry. In addition, to repeatedly discard the unused slurry would be both wasteful and costly. From the start I chose the second option because it would provide a constant supply of slurry ready for use.

EQUIPMENT REQUIRED FOR THE SLURRY

A suitable container for the slurry must combine strength and rigidity in a material that will not react with the colloidal silica. Recommended is stainless steel, which is widely used in industry but is too expensive for our simple operation. Plastic containers will work, but they must be rigid enough not to flex and cause dried slurry from the container walls to fall into the batch. A galvanized coating on a steel container reacts chemically to gel the colloidal silica; I found this out to my sorrow after selecting my first containers. Galvanized garbage cans were both available and affordable, however, and I was able to eliminate this gelling effect in the two forty-gallon cans I now use by coating the interior with polyester resin reinforced with fiberglass cloth. This lining also made the containers stronger and more rigid. For our campus operation, we use a plastic drum with a wall 3/16″ thick, obtained without charge from a major department/mail order store. The drum originally contained liquid floor polish.

For mixing the slurry, one has a choice of either a fixed blade in a continuously rotating container, or a stationary container using rotating mixer blades. Where constant dipping is required for industrial production, the fixed-blade mixer is preferred. The cost, however, placed it beyond our reach. For the mixer blades and shaft, stainless steel is also recommended, but again, the cost is high. My solution was to make a mixer from a steel rod (*not* stainless) one-half inch in diameter and 36″ long, and two 12″ pieces of 1″ angle iron. Leaving a section of the original angle in the center, a cut was made in the angle iron on each side; the two ends were then flattened out into opposite-angled blades like a propeller. As illustrated at the left, the two blades were brazed onto the steel shaft so that in relation to the direction of rotation, the bottom blade pushes up, and the upper blade, approximately 12″ above it, pushes down. The iron does rust, of course, but the deterioration seems very gradual; the blades will probably be worn away by abrasion before the shaft has to be replaced.

The next requirement is for some means to hold and propel the mixer. For my first slurry batch, this was accomplished with a Rockwell drill press which allowed the head (with chuck and motor) to swing out over the slurry container and to be cranked up and down. That arrangement worked fine until the 1/3 HP motor burned out; now I recommend nothing smaller than a ¾ HP motor to power a slurry mixer! Using that same drill press concept, my current mixer was begun with a simple angle-iron frame. Two pillow blocks in front hold the chuck and shaft, which are drill press replacement parts. A hinged mount for the ¾ HP-1725 rpm motor at the rear provides for adjusting belt tension. Part of a drill press speed-reduction system, the sliding pulley between drive shaft and motor reduces mixer rotation speed below 500 rpm. A speed higher than this will heat up the slurry, increase evaporation and foaming, and cause excessive splashing. When the slurry is to be used, the mixer shaft is dropped out of the chuck, and the entire unit swings to the right out of the way. After the lid has been removed, the mixer shaft is pushed to one side of the container to provide full access to the slurry. On campus we have adapted an old commercial dough mixer like those used in military kitchens to keep our slurry in suspension.

One more component is essential to maintain the slurry once it has been mixed—an automatic timer. It should have repeating sixty-minute cycles so that one-minute adjustments can be made. The exact

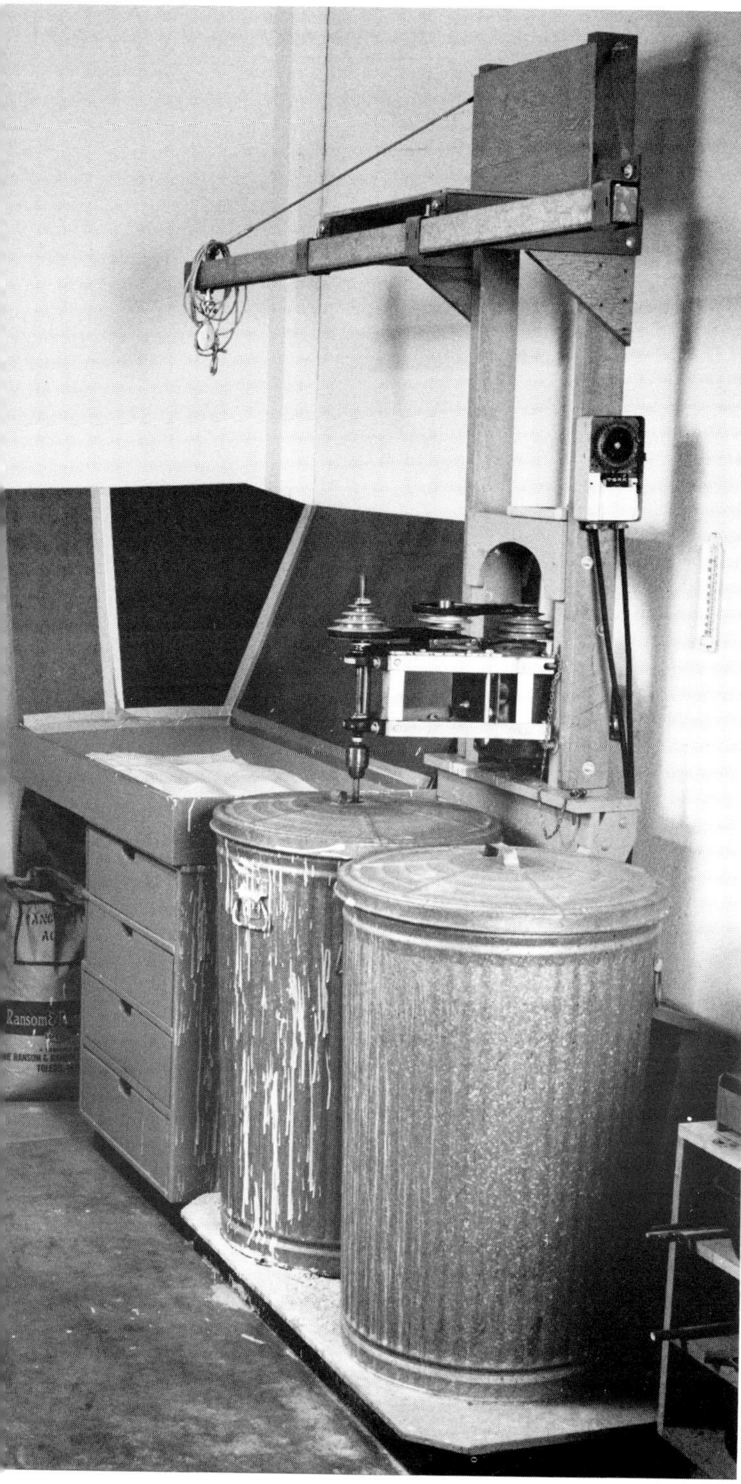

10.2 One corner of the studio is occupied by the slurry mixer and the stucco boxes. The two forty-gallon, galvanized garbage cans are lined with resin-impregnated fiberglass. The mixer is on an automatic timer, and stirs the slurry for three minutes every half-hour. The overhead track, designed for a sliding barn door, provides for moving large molds from the wetting dip to the slurry to the stucco box.

length and spacing of the on and off periods depend on how long the slurry can stand without settling. The size and angle of the mixer blades and the speed of rotation are key factors. My present slurry mixer runs for three minutes, then is off for twenty-seven minutes.

MIXING AND TESTING THE SLURRY

To prepare the slurry, the colloidal silica sol is first measured into the container. With the mixer blades rotating, the fused silica powder is then sifted into this milky liquid. A 15-gallon batch of slurry takes a little over 6 gallons of colloidal silica and about 135 pounds of fused silica powder. About three-fourths of the powder will be absorbed readily; as the mixture thickens, the silica must be added more and more slowly to avoid lumps. To allow for complete absorption of each dry particle, the batch should not be used until several hours after the last silica powder has been sifted in.

As the final step, a small amount of a wetting agent (such as Ultrawet 60L or Victawet-12) should be slowly added while the slurry is being stirred. About two tablespoons should suffice for a 15-gallon batch; too much will cause excessive foaming. This honey-like additive will reduce any tendency for the slurry to "crawl" on the wax pattern, leaving bare patches in the first coat. I always add some wetting agent to the pre-wet dip as well. This is the second container filled with only the colloidal silica sol. The sprued wax is dipped into this liquid to wet the previous mold coat just before immersing it again in the slurry.

To achieve shells of maximum strength, the slurry must be maintained at the proper viscosity. After some experience in making shell molds, one can tell simply by feel and appearance if the slurry is too thin. To be most accurate, however, the viscosity should be measured with a No. 5 Zahn cup viscosimeter and a stopwatch. The viscosimeter is nothing more than a stainless steel cup with a hole in the rounded bottom

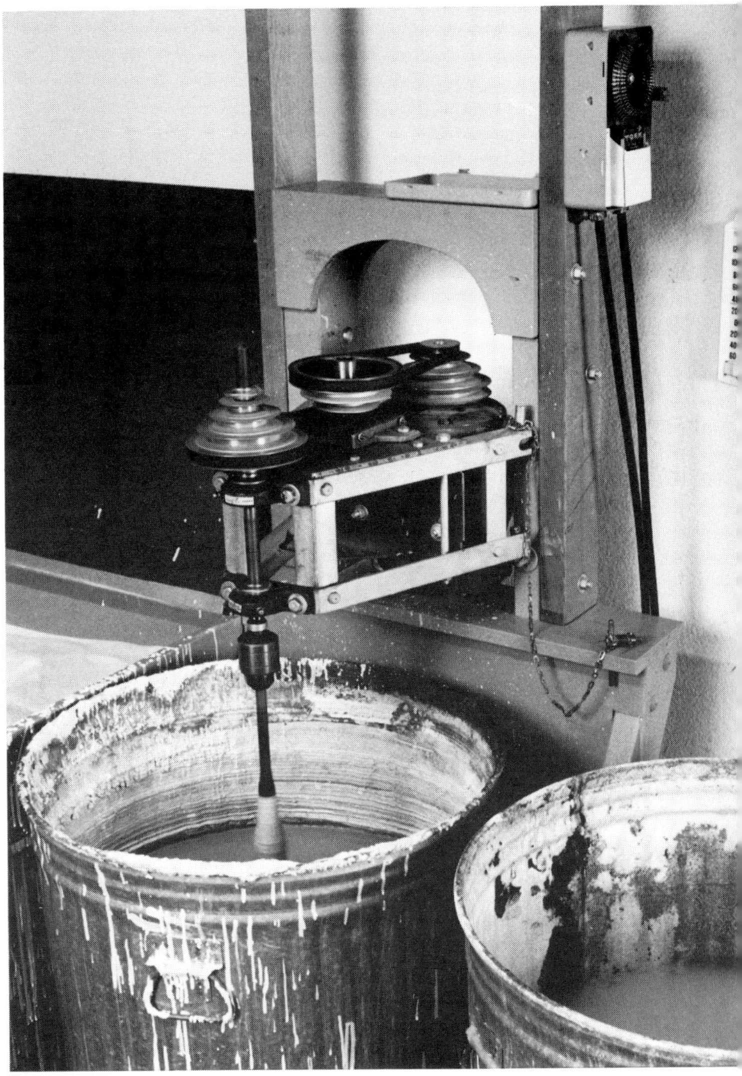

10.3 The slurry mixer is driven by a ¾ HP motor. Its speed of 1725 rpm has been reduced to one-fourth that by a pulley speed-reduction device made for a drillpress. A simple angle-iron frame holds the motor, central pulley, and drive shaft with chuck. With the mixer blade removed, the entire drive unit swings out of the way to the right so dipping can proceed unimpeded. The container on the right holds colloidal silica for wetting molds before they are immersed in the slurry.

and a long handle. After the cup is filled, one counts the number of seconds it takes for the slurry to drain out through the hole. For proper viscosity, this should be at least 30 seconds—some recommend 40 to 60 seconds. An inexpensive plastic viscosimeter purchased from a paint store can be timed to measure correct slurry viscosity. To check the density of the slurry, a specific gravity hydrometer for heavy liquids (graduated between 1.600 and 1.820) is used. The hydrometer, a fragile glass tube weighted with lead, will sink into the slurry until equilibrium is reached. The specific gravity should read between 1.75 and 1.80 on the vertical upper scale.

APPLYING THE STUCCO

There are several different ways for the stucco to be applied, but I recommend the simplest way—that of sifting or tossing the stucco by hand over the wet, slurry-coated work. All that is needed for this manual application is a box to contain each kind or grade of stucco being used. I have one box for fine fused silica stucco, and one for coarse. On campus we have added a third box for an even finer zircon flour for face coats. Constructed from ½″ plywood, each box is 2′ wide, 6′-8′ long, and 6″ deep. For convenience, the boxes are elevated to table height, with an apron in back to keep the tossed sand from falling behind.

In commercial foundries, continuous production demands more efficient procedures which require, in turn, more complicated equipment. One device provides for the stucco to "rain" down from above, the flow triggered by a foot pedal, while the work is being rotated in both hands. Another more complex piece of equipment is the *fluidized bed*—a deep container in which air, forced upward through a porous floor, suspends the stucco particles several inches above. The work can be dipped and rotated in this layer of floating particles as if it were liquid. However, the time and cost involved in constructing, installing, and maintaining a fluidized bed generally outweigh its advantages for the intermittent use typical of the studio/classroom and the sculptor's studio.

DRYING AND REINFORCING MOLDS

Forced air is the only practical way to accelerate the drying of mold coats, since heat would expand the wax and crack the shell. Variable-speed, portable fans can circulate the air throughout the drying area or be focused directly on the molds. Compressed air can be channeled through rubber or plastic tubing into deep cavities to dry them faster. Too strong a breeze on the first or second coat, however, may chill and shrink the wax so that the thin mold coating will push up away from the surface, resulting in ridge-like scabs on the metal casting. Only if the mold room is kept at a constant temperature and humidity can shells of maximum strength be produced consistently.

To be most effective, reinforcement materials should be embedded in the mold wall. The application of Refrasil woven fiberglass tape has been described in Chapter 3. Developed to withstand temperatures up to 3000°F., the tape goes on readily because, unlike regular fiberglass, it absorbs the slurry. Widely recommended for reinforcing large molds is stainless steel wire, which should be wrapped around the mold about midway through the sequence of coatings. Soft iron wire, however, can be substituted for reinforcing small molds. For flat panels or reliefs which tend to split around the perimeter, U-clips made from 8d finish nails or 3/32″ welding rod can be spaced at intervals around the edges to prevent mold separation. If cracks occur in the burnout, the mold can be wrapped with wire, then kiln mortar can be spread over cracks and wire alike. The mortar must be baked dry before the mold is placed in the preheat kiln, where it will be completely hardened prior to pouring.

11 BUILDING EQUIPMENT FOR BURNOUT AND PREHEATING

A major difference between ceramic shell and solid investment moldmaking is the method of burnout. As described in Chapter Four, ceramic shell molds must be *shock heated,* that is, inserted into a kiln hot enough to melt the surface of the wax almost instantly through the mold wall. Besides heating capacity, a burnout kiln for ceramic shell molds must provide a safe and efficient way to (a) get the molds in and out, and (b) evacuate the melted wax and collect it for reuse.

Ceramic kilns do not work well for this purpose because they are built with doors either on the side or on top. My earliest shell molds were burned out in a side-loading ceramic kiln which had elements in the door; every time I set a mold inside the hot kiln, my right arm got toasted! Lowering a mold into a preheated top-loading kiln is even more difficult. To minimize both the heat loss and the heat hazard during mold insertion, the best type of kiln seemed to be one which, after preheating, could be raised, allowing the mold to be placed under it, and then lowered. My first test version of such a *sleeve* kiln was made from 30-gallon garbage can. The interior lining was cast from a refractory mix around a Styrofoam core. Electric elements were used for heating. Metal loops attached to the top provided for a rod to be inserted for lifting. The concept worked beautifully, but the kiln interior was too small.

The next move was to make a larger kiln of the same type using a 55-gallon steel drum, as shown to the right. The bottom was first removed with a cutting torch, leaving a lip all around to support the firebrick. A disc of metal lath was suspended inside the top of the drum; with the drum inverted, a castable refractory was poured in to make a reinforced ceiling 3″ thick. The four vertical strap steel hangers were then bolted into place and welded together at the top.

The walls of the drum were lined with soft 2300°F. firebrick. The long edges of each brick were beveled in a jig so that each circular course of vertical bricks was locked tightly within the drum. Notches were also pre-cut for the elements to fit into. The firebrick courses were laid with kiln mortar from the bottom lip of the drum upward, the top course fitting tightly against the cast ceiling.

The electrical circuitry for the kiln was custom-designed by a major kiln manufacturer based upon the power available in our foundry. The six elements, two circuit-breakers, and everything needed to connect them were provided for our installation. A galvanized steel box was made to cover the electrical panel, and two 25-foot heavy-duty cords with plugs completed the project. The kiln cost less than $200 to build in 1971.

The same sleeve kiln is also used for preheating ceramic shell molds just before they are to be poured, as shown in Chapter Five.

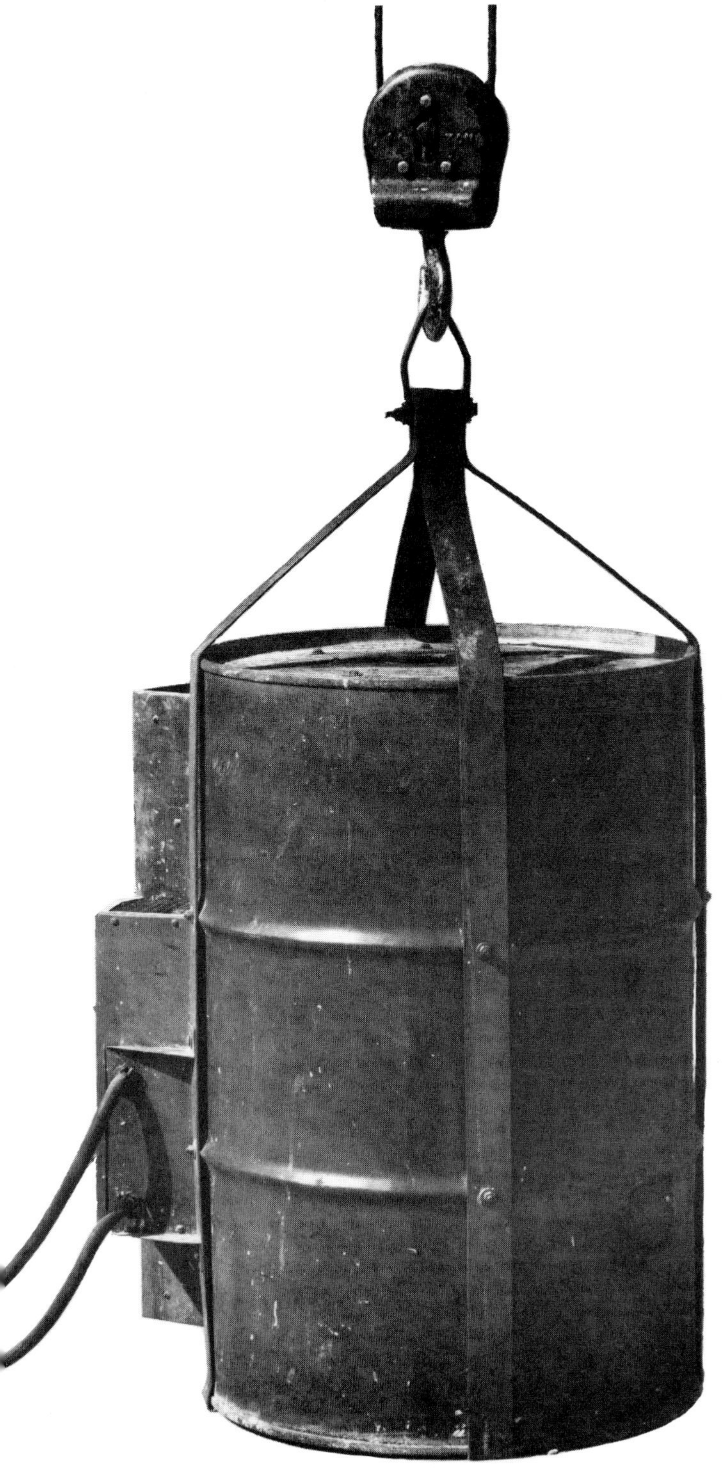

11.1 Made from a 55-gallon steel drum, this sleeve kiln is used both to dewax and fire ceramic shell molds and to preheat them just prior to pouring. The ceiling is a solid-cast refractory, with beveled, soft firebrick set vertically forming the walls. Six coiled elements on two 30-amp circuits provide rapid heating. The kiln has now been in use fourteen years.

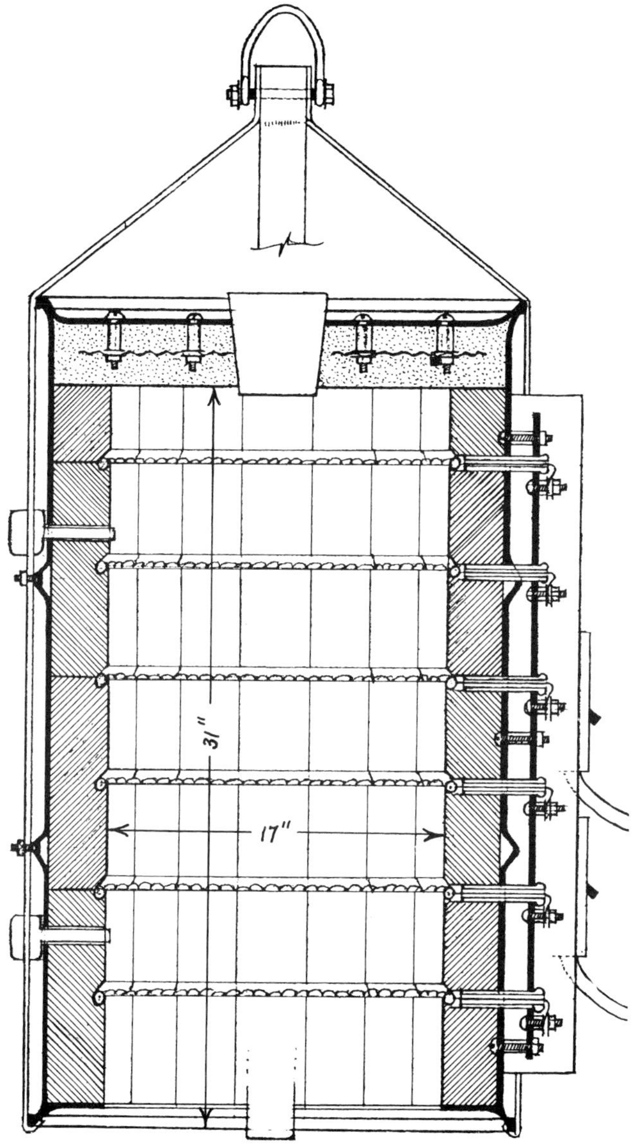

An efficient burnout system for ceramic shell molds must also provide a way to rapidly drain and collect for reuse the melted wax from the kiln interior. Most widely used to perform this function is the *drop-bottom* furnace in which the kiln floor is replaced by an open steel grate. The flaming wax drops through this grate into a shallow pan of water. Unless the kiln is considerably elevated and the water continuously cooled, however, the wax floating on the water's surface may continue to burn vigorously. The fluid wax may also mix with the water and have to be discarded.

The inherent difficulty in using a steel grate is that each time it becomes red-hot, the metal surface is converted into a thick layer of oxidation. This layer subsequently flakes away, and before long the entire structure has so deteriorated that it must be replaced. Is there any material, I wondered, that can be brought repeatedly to a red heat without deterioration? The answer was immediately at hand—the shell mold material itself! My first version of a drain base of cast fused silica slabs was constructed in the mid-sixties to set inside a side-loading kiln. When the first sleeve kiln was completed, that drain basin was encased in firebrick to serve as the kiln base for another ten years. The current version is lightweight and portable, and the insulation underneath allows the wax to cool below its ignition point before flowing into the collecting vessel.

For those who might wish to build such a drain base themselves, the steps are outlined below:

1. The parts for the frame were first cut from 1½″ angle iron, and the corner diagonals from ⅛″x1½″ flat steel. The metal frame was then welded together.

2. The assembled structure was placed upside-down on a table, and the outside edges were sealed with clay. A 14″ square of ¾″ plywood covered with transparent plastic wrap was centered within the 24″ angle-iron square, sealed with clay on all four sides, and weighted down. Into the surrounding space slurry was poured to a depth of ⅛″. Fine stucco was then sifted onto the wet slurry until it no longer sank in. Alternating layers of slurry and fine sand were applied until the thickness reached ⅝″. When the cast slurry was dry, the plywood center form was removed, leaving a border 5″ wide around the top of the drain base.

3. Triangular cardboard patterns were cut to form a four-sided pyramid rising up out of the square center opening. Forms were prepared by nailing wooden strips ⅝″ thick onto plywood which had been sealed with plastic wrap. Four triangular slabs were then cast in these forms by alternately pouring in slurry and sifting fine stucco. When dry, the edges of the triangles were beveled with a file. Using slurry as one might use slip in slab clay construction, the four panels were assembled into a pyramid inside the square center opening. With fine stucco added, the joints could be smoothed with the fingers.

4. A hole ¾″ in diameter was drilled through the point of the pyramid. Over this hole a one-inch iron pipe elbow was fitted. Strips of Refrasil tape were dipped in slurry and wrapped over the elbow to fasten it securely.

5. After making cardboard patterns, four triangular corner braces were cast in the same manner. These slabs were fitted and attached from the apex of the pyramid diagonally into each angle-iron leg.

6. The entire unit was then returned to an upright position, and patterns were cut for the vertical panels that fit across inside the basin. These triangular slabs were cast and assembled in grid fashion to create a level, horizontal floor with notches at the base through which the wax could drain.

7. Finishing included: firing with a torch all cast parts that would not be exposed to the kiln heat; attaching a U-bolt for the drain pipe; cementing rigid Fiberfrax an inch thick onto the bottom of the drain basin; and painting the metal frame.

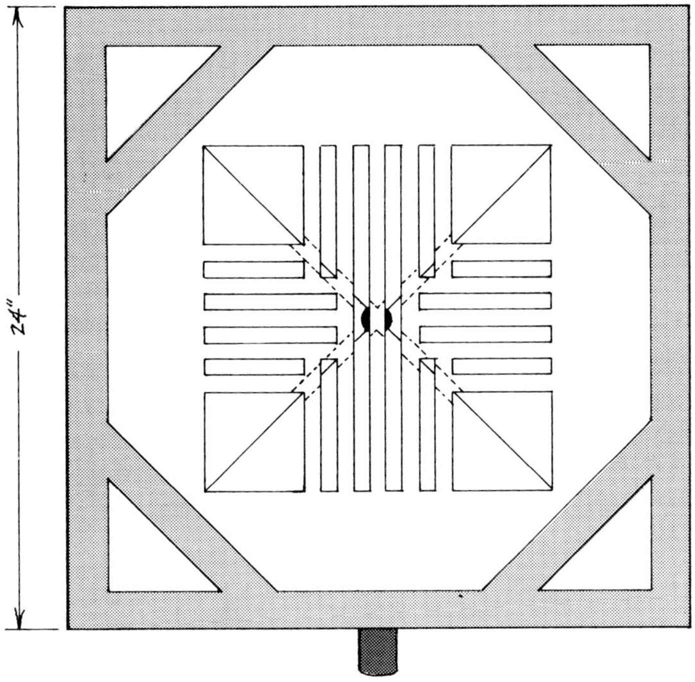

11.2 This drain base, made from cast slabs of the same fused silica slurry used for ceramic shell molds, can be brought to a red-heat and cooled suddenly and repeatedly without damage. The outer square of the top surface was first cast into the inverted angle-iron frame. The pyramidal basin was then constructed of ⅝"-thick precast slabs, with a one-inch pipe elbow attached with strips of Refrasil tape. The bottom is insulated with one-inch-thick Fiberfrax board.

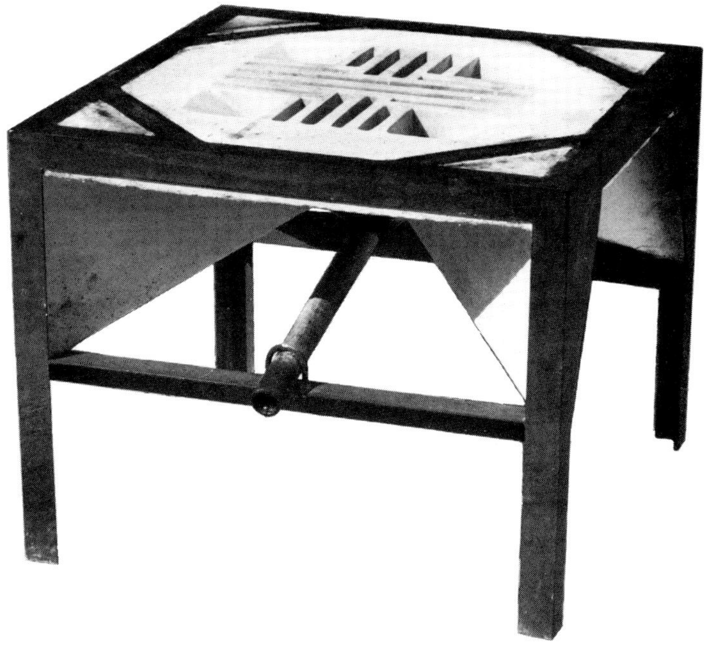

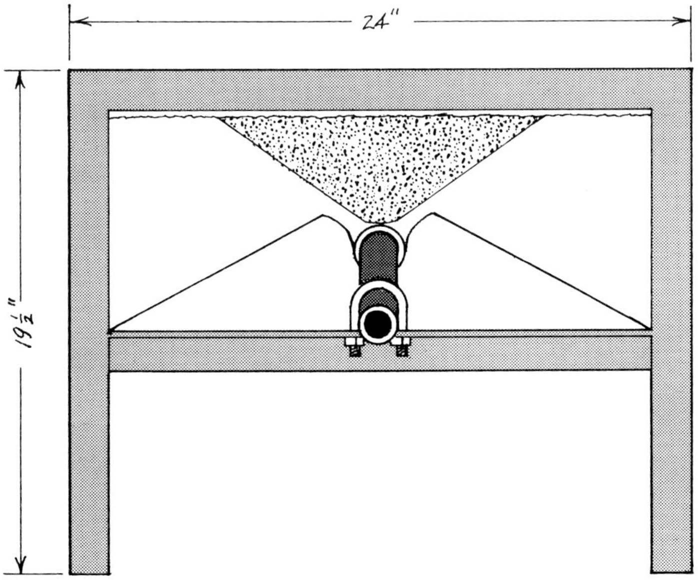

12

STUDIO PROCEDURES FOR WAX AND METAL

This chapter is intended to supplement the limited coverage of tools and techniques included in the brief captions and text of Part One. More detailed procedures and a wider range of optional approaches to the various steps in the process of ceramic shell bronze casting follow.

SELECTING AND MODIFYING WAXES

Wax is an extremely versatile material. It can be modeled, carved, cast, spread, fused, assembled as rigid components, or used to coat forms dipped into it. The method one chooses for shaping the wax, however, depends on the properties of the wax to be used, especially its hardness or pliability. A wax for modeling, for example, must have a plasticity when warmed that resembles clay; a wax for carving must be much harder without being brittle; and a wax for casting must be tough and yet elastic. Waxes must also burn cleanly from the mold, be rigid enough to withstand handling and not collapse in moderate temperatures, exhibit minimum shrinkage, and be low in cost.

The earliest sculptors used natural beeswax; wax formulas since the Renaissance usually begin with beeswax as a base. However, a twentieth-century synthetic byproduct of the petroleum industry—microcrystalline wax—has proven to be even more pliable over a wider range of temperatures than natural waxes because of its smaller crystalline structure. Much less expensive than beeswax, it can be purchased in 11-pound slabs from major oil companies directly or from art supply houses. Microcrystalline wax comes in varying hardnesses and colors, but least expensive is the unmodified pale yellow which can be readily colored (to be more like the eventual bronze) with a package of powdered household dye.

As pointed out in Chapter 1, microcrystalline wax by itself is the most pliable for modeling, but thin sections must be reinforced with toothpicks, bamboo skewers, or other combustible supports. When new, this wax may not always be fully plastic, but as it is collected in the burnout and reused, its plasticity increases like an aged clay body. In warm temperatures microcrystalline wax will soften and sag even more, and should be stiffened with paraffine, which can be added up to as much as one-third of the total weight.

For carving or turning wax forms, higher proportions of paraffine and other hard waxes such as carnauba must be used. It is simpler to buy blocks of hard carving waxes from jewelry suppliers than to mix them. Casting waxes need a tough elasticity to prevent distortion of the cast during removal from the mold. The addition of rosin (ten to twenty percent) to the wax mixture will provide the extra strength needed for crisp wax castings. The rosin should always be melted first before the other waxes are added.

POURING AND SHAPING WAX

In Chapter 1 wax sheets were made by pouring melted wax into shallow trays of heavy-duty aluminum foil. This is only one of several different ways to create wax sheets. To get sheets of the most even thickness, melted wax should be poured onto a level marble slab that has been lubricated with diluted detergent or green soap and allowed to dry. The wax is confined by strips of angle iron, which can also be used to fence in wax sheets poured onto a piece of damp, heavy-weight canvas fabric. Another approach is to use a flat plaster mold recessed to the depth desired for the wax sheet. The mold must always be soaked in water until saturated and the surface water absorbed before pouring in the wax, which will stick to dry plaster.

Some cautions are in order concerning melting and pouring wax into any type of mold. If previously-melted wax, solidified in a metal pan or bucket, is to be remelted, always melt a channel first alongside the pan from bottom to top of the wax mass. Otherwise, the melting, expanding wax at the bottom may erupt with explosive force, splattering hot wax over the surrounding area. Nothing seems more painful than a hot wax burn because it clings to the skin; the only relief is to dunk the coated limb into ice water.

Plaster piece-molds must also be saturated with water before wax is poured into them. A light coating of mold release should be sprayed or brushed onto flexible molds. The wax should be only slightly above its melting point when poured. For hollow forms, a slush-casting technique is used—the wax is allowed to cool for just a few seconds, leaving only a thin coating on the mold wall before it is poured back into the container. By successively filling the mold with wax and emptying it, the full wax thickness (which will be the final bronze thickness) is achieved.

Hollow spherical and ovoid wax forms can be created by dipping a water-filled balloon into melted wax. More detailed shapes modeled out of water-based clay can also be coated with wax by repeated dipping; the clay inside can then be dissolved by soaking the entire form in cold water.

The strangest and most unique wax shapes can be produced by gently pouring melted wax into water. The rate at which the wax is poured and the temperature of both the wax and the water all affect the formation of the wrinkled, drapery-like wax sheets that grow on the water's surface. Slowly lowering a container of melted wax down under the water can create flame-like tongues of wax which solidify as they ascend toward the surface.

OTHER TOOLS FOR WAX WORKING

Although my favorite tools for working with wax happen to be a grapefruit knife and an alcohol lamp, the more traditional tools are an iron rod with the end forged into a spatula, and a Bunsen burner. The rod is supported so that the flattened tip is kept hot in the flame and ready for use. A more sophisticated device can be made from an electric soldering iron with interchangeable forged tips and a rheostat to vary the current. More expensive manufactured wax-working tools with adjustable heat controls can be purchased from jewelry suppliers.

Wax sheets and sprues often need to be warmed and softened for bending or modeling. Most practical for this purpose is a clamp-on metal reflector with an infrared bulb. Also effective is a metal can lined with aluminum foil with a single light bulb suspended inside—a kind of reflector oven.

CASTING YOUR OWN SPRUES

For sprueing wax patterns, extruded wax rods of varying thicknesses, both solid and hollow, can be purchased. Solid round and square rods come in 24″ lengths and range from ¼″ up to ¾″ in diameter in ⅛″ increments; above ¾″ they are usually hollow. Extruded sprues are generally made of waxes that are a different color and softer than most pattern waxes so the sprues will melt first. Although purchasing sprues is cost-effective for the commercial foundry, the cost may be significant for the student and individual sculptor. I have found it highly economical to cast my own sprues, especially since I use repeatedly the same wax reclaimed each time from the burnout. Two-piece sprue molds can be readily cast from plaster by using shellacked wooden dowels of different diameters for the original patterns.

OTHER TYPES OF POURING CUPS

The pouring cone or cup fashioned from sheet wax and described in Chapter 2 evolved from my desire to handle the sprued wax entirely by the cone, if possible. Touching the wax anywhere else during investment inevitably thins and weakens the shell; extra coats must be added at those contact points to equalize the mold thickness. Wax cones must be strong and securely attached to withstand manipulation of the sprued wax in the slurry and stucco. Waxed paper or foam plastic cups are traditionally used for solid-investment casting, but if used for ceramic shell casting, the sprued wax could never be grasped by or balanced upon the pouring cone.

In production foundries, an eye-bolt or a pipe tee may be cast into a solid wax cone made from a mold. The bolt or pipe functions first as a handle for dipping and stuccoing, then as a means for hanging the work upon a rack to dry between mold coats. When the mold has been completed, the metal handle is unscrewed from the pipe flange or washer and nut embedded in the wax cone. These small metal parts are then retrieved in the burnout.

CASTING DEFECTS AND REMEDIES

Lest the reader assume that a perfect casting comes from every mold, listed below are some of the most common defects that can occur in ceramic shell casting, and what should be done to avoid them.

1. *Misrun,* also referred to as *non-fill,* simply means that the metal solidified before the entire mold cavity was filled. This can be corrected by (a) pouring the metal at a higher temperature; (b) preheating the molds longer so they will be hotter when poured; and (c) delaying removal of the preheat kiln so that molds will not cool so much before pouring.

2. *Flashings* are thin, fin-like metal projections on the surface of the casting, caused by metal flowing into a crack in the mold wall. Cracks can be prevented by (a) reducing fluctuations in mold room temperature and relative humidity; (b) preheating the burnout kiln to a higher temperature before inserting molds; (c) making thick pattern sections and main sprues hollow; (d) reinforcing molds with fiberglass tape and wire prior to burnout; and (e) handling molds more carefully when removing the mold cap before burnout.

3. *Metal breakthrough* produces tiny bubbles of excess metal inside sharp corners, crevices, and holes. It is caused by incomplete slurry penetration into these recessed areas due to air bubbles or crawling of the first slurry coat. This defect can be corrected by (a) adding more wetting agent to the slurry to eliminate crawling; (b) blowing harder to remove bubbles in all corners and openings during the first slurry coat; (c) painting the face coat onto a heavily-textured wax; and (d) cleaning patterns using extruded sprues with a spray of isopropyl alcohol (allow to dry before dipping).

4. *Shrinkage* cavities occur in a metal casting when a section has an insufficient supply of fluid metal as it cools. Such defects can be corrected by (a) attaching larger sprues, making sure that every sprue is as thick as the section it feeds; and (b) pouring the metal at as low a temperature as possible to fill the mold.

5. *Inclusions* are unwanted particles of foreign material, usually from the mold, that are embedded in the metal casting. This problem can be avoided by (a) applying a final coat of slurry without stucco to

adhere all stucco grains to the mold; (b) filling solidly every joint in the sprued wax to prevent the formation of thin projections that could be broken off inside the mold by the metal flow; (c) shaking out or vacuuming the mold after the burnout to remove any loose particles; and (d) exercising care in lowering the preheat kiln over upright molds.

6. *Cold shuts* are hairline crevices on the casting where two separate metal flows have come together but have not fused. This defect can be avoided by (a) pouring the metal at a higher temperature; (b) preheating the mold to a higher temperature; and (c) planning the sprueing system more carefully so that either more feeders are used or the system is changed from direct to indirect.

The most calamitous defect of all is to have the mold crack open while it is being poured. Strong shells result from careful attention to each stage of the process, including maintenance of the slurry at the proper viscosity and the mold room at adequate temperature and humidity levels. However, only from experience does one learn how to plan the most efficient sprueing systems, how to determine just how thick a mold should be, and when it needs to be reinforced.

REPAIRING/JOINING CASTINGS

If one of the misfortunes from the preceding list does occur in the casting process, that defect must be repaired. Positive defects involving excess metal are easiest to remedy, since they require only the removal of the projecting bronze. After flashings and tiny bubbles of metal have been cut off with a cold chisel or hacksaw, the surface of the casting can then be filed and textured by the usual chasing techniques.

The negative defects—the cavities or absence of metal—are the ones that challenge us. Small nicks and gouges can sometimes be eliminated by using a jeweler's chasing hammer; properly annealed bronze is soft enough to be moved sideways by repeated blows until the indentation is filled. Irregular shrinkage cavities are more effectively filled by welding. Although an oxy-acetylene torch can be used with flux and a sprue from the casting, heliarc welding provides the strongest and cleanest repair. Portions of a casting which are missing due to non-fill can be modeled again in wax right on the casting; when sprued, invested, and cast, the bronze patch will fit precisely for welding.

If a large sculpture created whole in wax is to be cut into sections for casting, the cuts should be deliberately made irregular so the edges will fit together in only one way. Where two components are to be joined, one side can have a lip added that will fit inside the other—like the lip on a teapot lid. Another way to join separate castings into a larger sculpture is to cast bronze pegs atop one section, drill matching holes in the other section, then fit and weld the two sections together.

PRECAUTIONS IN PATINATION

Even though no acids are being used, great care must be taken in handling all chemicals used for patinas. Wash hands thoroughly with soap and water after all contact with them; better yet, wear rubber gloves during preparation and application of all patina solutions. Containers and spray bottles of glass and plastic should be clearly labeled. Brushing and spraying chemicals on hot metal should be done only in well-ventilated areas, and a breather mask is strongly advised.

In case you might expect a given recipe to produce a predictable and consistent result, be forewarned that the color may not turn out as anticipated because of the great number of variables in patination—the alloy, the metal structure, the chemical solution, the temperature of the bronze, etc., Successful patinas take much patience and experimentation.

EPILOGUE: EXPLORATIONS, MODULES AND MULTIPLES

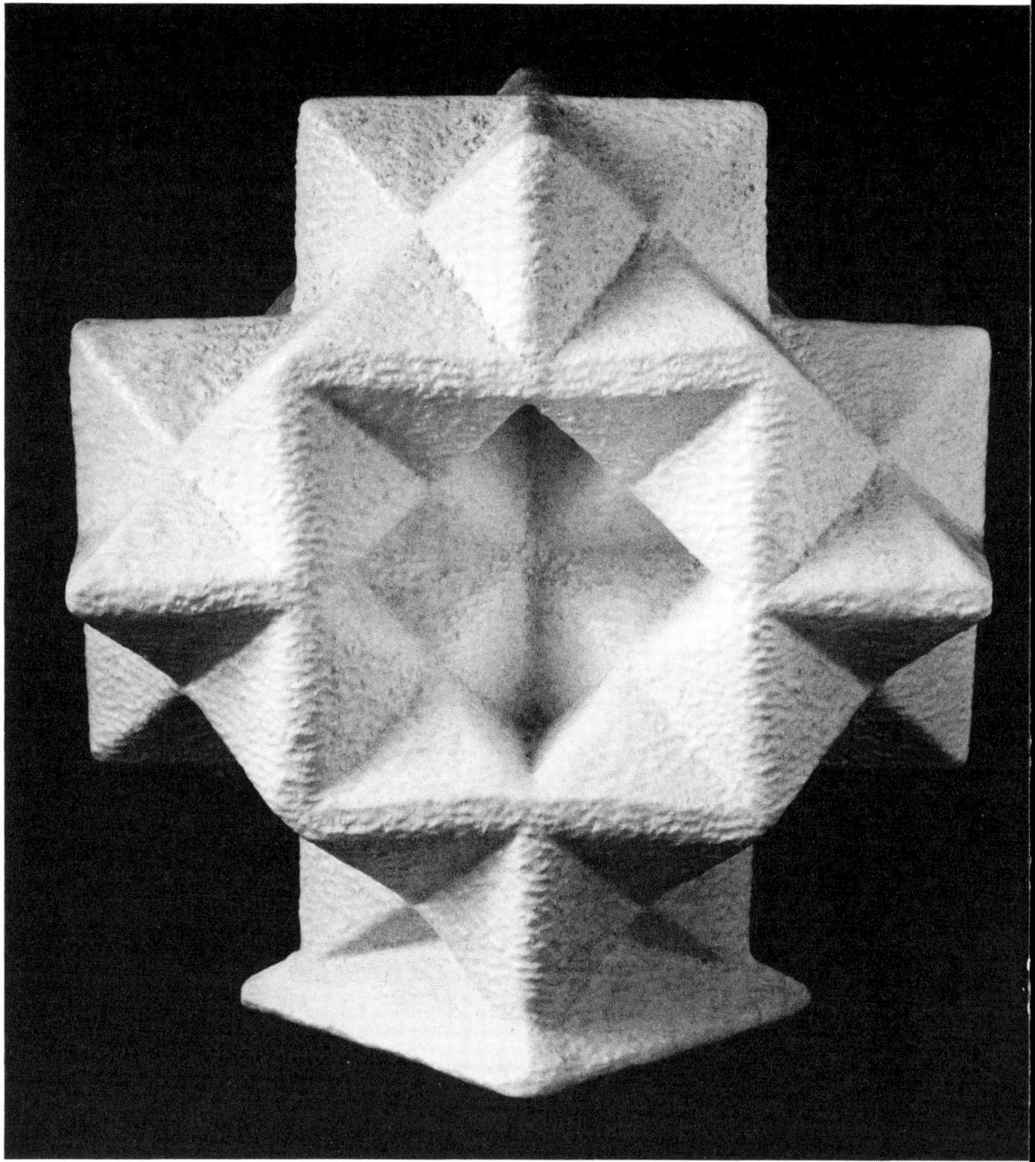

Howard Hitchcock, **White Starcross,** 1977. ceramic; 13¼"x12¾"x8". Collection of the artist. This finished sculpture is made from the ceramic shell mold material that is normally destroyed in bronze casting. The original form was constructed of wax, and a thin mold was built around it as if it were to be cast in bronze. Burnout and firing left a hollow shell; slurry was poured in and out of it several times, adding more coats to the *inside.* When the wall was thick enough to exist as a ceramic sculpture, the work was given a final firing.

Many sculpture processes are unpleasant to work with—dirty or messy, dusty, or noisy, requiring protection for the eyes, ears, hands, or lungs. For me, making ceramic shell molds has always been a pleasurable experience, aesthetically satisfying and without physical discomfort. I enjoy building sturdy wax structures, dipping them into the cool, viscous slurry, and sprinkling them with sparkling white sands that turn them into shapes resembling sugar-coated confectioners' fantasies. How many times someone admiring a completed mold has exclaimed, "Why not leave it that way—it's beautiful as it is!" But that form is not what was intended, I explain, and all of these linear components will have to be removed later to reveal the true form. What if I were to make a mold, I thought, which was not just an expendable vessel in which other materials were transformed, but was intended to stay as a matte white, textured form?

EXPLORATIONS

To explore the ceramic shell mold material as a sculptural material in its own right, I created a series of ceramic shell sculptures, including the one at the left, from fused silica slurry. One of the most amazing characteristics of the material used in *White Starcross* is the ability of the moist slurry to adhere securely and permanently to the *fired* shell without shrinkage. With clay this would be impossible! Keeping the outer mold coats to a minimum to retain as much definition of the wax surface as possible, I built up sufficient strength and thickness in the sculpture wall *after* the burnout by adding slurry coats to the inside of the fired shell.

Another unique property of the fused silica shell verified by my experimentation is its ability, as a ceramic material, to be glazed as well as fired by insertion into a hot kiln—a process we know as *raku.* The ceramic shell is the ultimate raku material, since it can be heated or cooled suddenly without fracturing, a quality that has never been achieved to such a degree with traditional raku clay bodies. Some of my ceramic shell sculptures were glazed with raku glazes, using the same sleeve kiln designed for burnout and preheating of bronze molds.

The fused silica slurry can also be cast into slabs, as described in Chapter 11, to construct sculptural forms as well as functional items like the drain base. When dried, the slabs can be sawed, filed, and sanded prior to assembly with slurry. High-temperature fiberglass tape can be added to reinforce joints if needed. After firing, the pure white, matte surface closely resembles porcelain.

These distinctive physical properties of fused silica comprising ceramic shell molds provide a challenge for further exploration into other ways that it can be effectively used. One potential application that I have yet to explore is the use of ceramic shell molds for casting glass, since the shell can be molded into any shape and can be heated to the same temperature as the molten glass without structural damage or dimensional change.

MODULES AND MULTIPLES

A bronze cast from a single, directly-constructed wax original is referred to as a *unique* bronze, which means that no duplicates exist or will be made. If some misfortune occurs during the casting process, such a one-of-a-kind work would be lost. To submit for casting a wax that cannot be duplicated is too great a risk for some sculptors; it requires absolute confidence in the founder. Commercial foundries cannot afford to lose a client's work through error or accident; their first step is to make a flexible mold of the original sculpture so that it can be duplicated in wax. Since I cast my own bronzes, I find the qualities of form and surface which result from directly shaping the wax worth the risk.

Besides eliminating the risk of losing unique works, many sculptors prefer for economic reasons to make *editions* of their bronzes, duplicating the

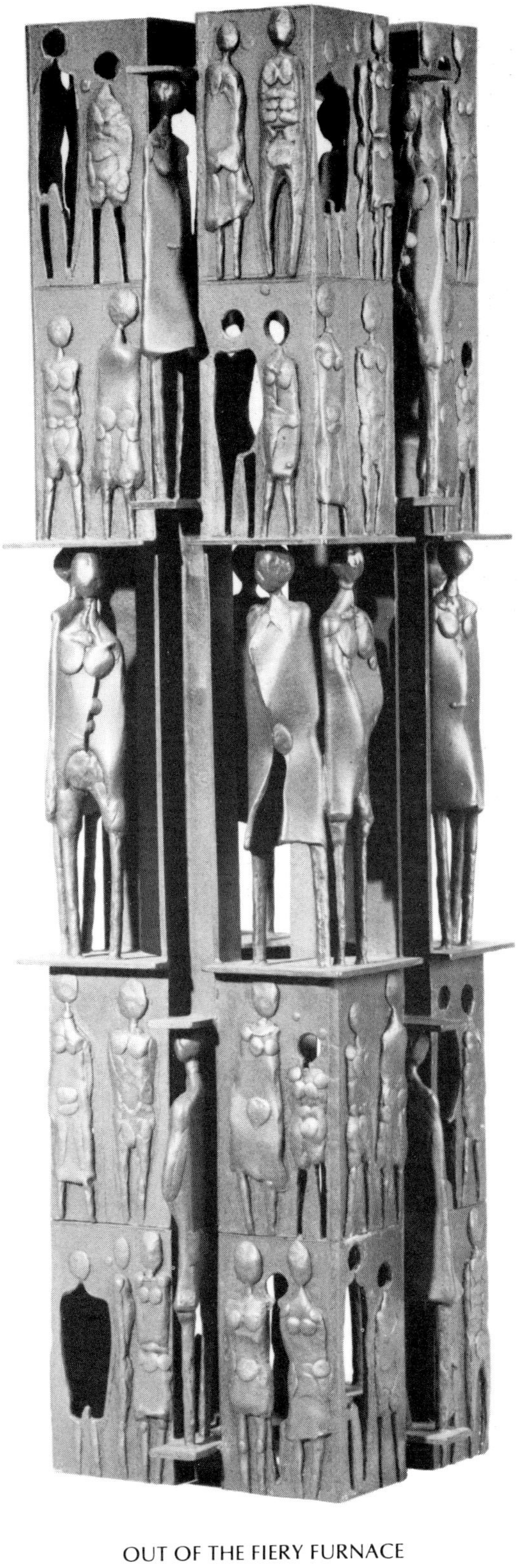

Howard Hitchcock, **Princeton Series #4: Tower,** 1981, bronze; 33″x9¾″x9¼″. Collection of the artist. This work and the relief to the right are part of a series of bronzes made from modular wax components which were assembled in ten different configurations. One modular unit was a 12″x12″ relief, the other a group of four free-standing figures.

Howard Hitchcock, **Princeton Series #3: Populace,** 1981, ▶ bronze; 31″x43⅞″x1½″. Collection of the artist. Modular wax components were combined with sheet wax strips and boxes to create this relief. The wax modules, cast in flexible molds, were assembled in subsections that fit the burnout kiln. After casting, the bronze sections were either welded or bolted together to form the whole.

original work a designated number of times. To create an edition, a reusable mold of the sculpture must first be made, a mold in which wax duplicates can be cast. Each wax casting must then be sprued, invested in an expendable mold, and cast individually in bronze as described in Part One. An edition presupposes an original form from which a mold can be taken. A plaster or flexible mold can be made with relative ease from monolithic forms with few openings. However, as I have sought to build ceramic shell molds of ever-increasing strength and efficiency during the last two decades, my sculpture has evolved into more linear and open forms. To make reuseable molds of these forms is difficult if not impossible, at least not without cutting the sculpture into pieces and making a separate mold of each piece. A duplicate of the whole wax sculpture could then be made only by assembling the wax parts cast from these separate molds.

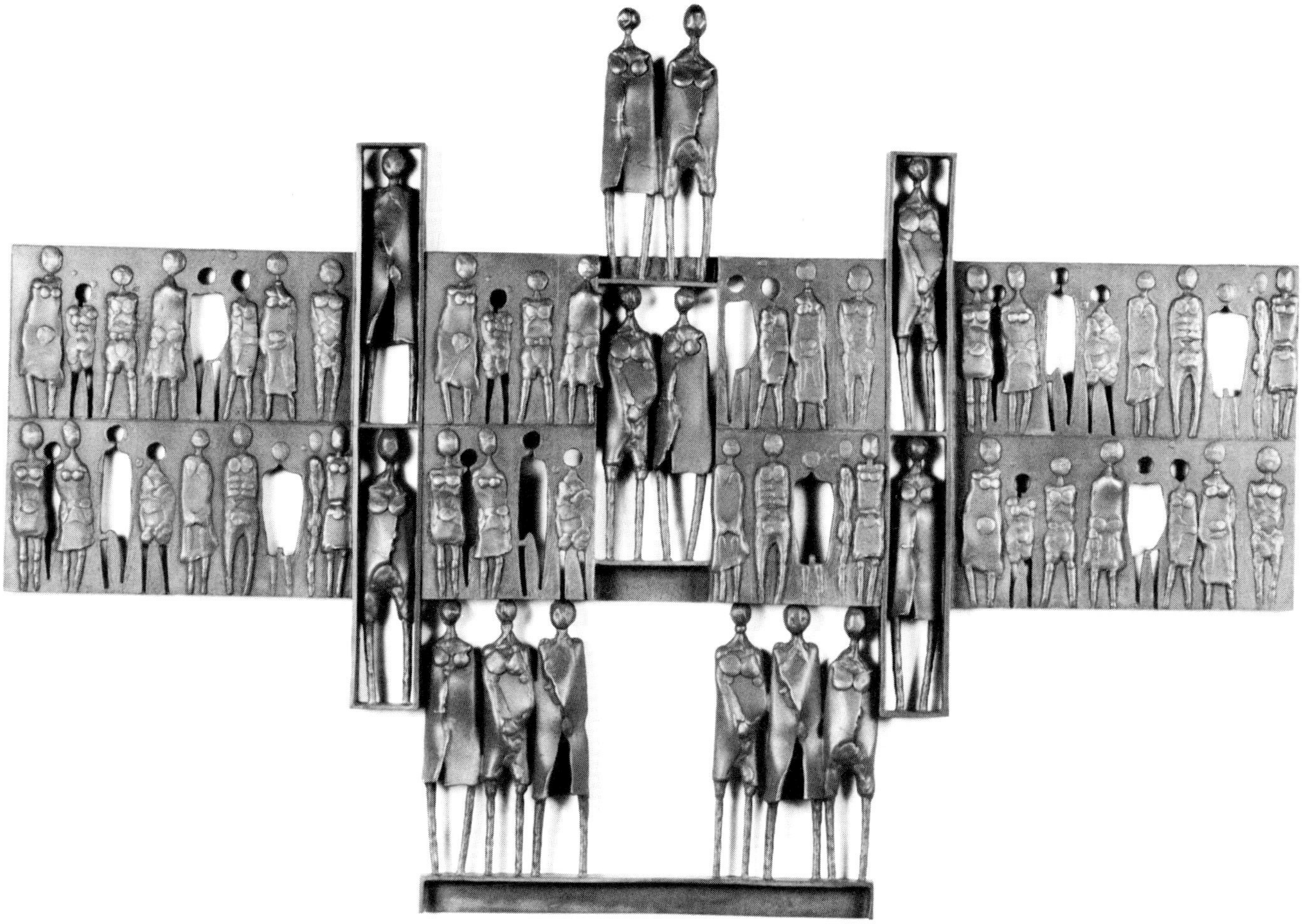

Since the assembly of wax parts seemed unavoidable, I began casting the parts *first,* as modular units that could be joined into not just one whole, but into many different configurations. My first module was a six-inch figure made from a two-piece plaster mold. Various numbers of these wax figures were combined with box-like components constructed from sheet wax to create an entire series of small bronzes. Each time I produced a mold for a segment of a newly-conceived work, I had the potential for a new module. Extra waxes from a variety of molds were left sitting around the studio. The urge to join them into new arrangements eventually became irresistible, and other series were generated.

Since I have been working with forms that are too complex to be duplicated whole for an edition, my approach to *multiples* has thus been through the use of *modules,* which results in a series rather than an edition. Unlike an edition, however, each sculpture is different and can be considered a unique bronze, even though its modular components are identical with other works in the series. Two bronzes from my most recent series are illustrated at the left and above. By using two different modular units cast from flexible molds, one a relief and the other freestanding, many more variations were made possible.

A FINAL NOTE

In visits to commercial and campus sculpture foundries across the nation, I have encountered a wide array of systems, many improvised, for ceramic shell casting. Countless other sculptors have obviously been intrigued, as I have been, by this process, and have figured out their own unique ways to make it work. I would welcome information about other approaches and methods for meeting the technical demands of ceramic shell casting in a campus or studio setting.

GLOSSARY OF TECHNICAL TERMS

Anneal. To heat and then cool metal to relieve internal stress, making it softer and more stretchable. Copper is annealed by quenching but steel is hardened.

Bronze. An alloy composed mostly of copper, plus a smaller amount of tin (traditionally 10 percent) or other metals.

Burnout. To eliminate the wax from the mold. Solid molds are slowly baked, while ceramic shell molds are shock-heated and fired.

Chase. To work the surface of the metal with various tools to achieve the desired finish.

Colloidal silica sol. The liquid binder which is combined with fused silica to make ceramic shell molds. A colloid is a liquid made up of very small insoluble particles (silica) that remain in suspension in a surrounding liquid (water). Every drop of colloidal silica contains millions of tiny silica particles. A sol is simply a colloidal dispersion in a liquid.

Crucible. A container for melting metals, made of a material such as graphite that can withstand intense heat.

Devest or *divest.* To remove the mold material from the metal casting.

Extruded. To be forced through a die to give a certain shape, the method by which wax sprues are produced.

Fiberfrax.® A lightweight, "space-age" ceramic insulation material made by the Carborundum Co.

Fused silica. Silica in the form of quartz which is melted and then ground into powder or granules; almost all of the shrinkage occurs in this first melting.

Gates, gating. An alternate term used for sprues and runners; more specifically, the exact point where the sprue attaches to the wax pattern and allows the metal to enter the main mold cavity.

Hydrometer. An instrument for measuring the specific gravity of liquids.

Intaglio mold. A negative shape or cavity carved into a hard material such as stone.

Invest. To coat or surround a sprued wax pattern with mold material; to build a mold.

Microcrystalline wax. A synthetic wax derived from petroleum with a crystalline structure that can be seen only with a microscope.

Patina. A color change on metal surfaces caused either by natural oxidation or the deliberate application of chemicals.

Pyrometer. An instrument for measuring unusually high temperatures.

Quench. To cool completely by plunging into water.

Refractory. Resistant to heat; a material used for molds or furnace linings that can withstand high temperatures.

Refrasil®. A fiberglass material capable of withstanding temperatures several times higher than the melting point of ordinary fiberglass; produced by HITCO, Gardena, CA.

Risers. Another term used for vents, the hollow channels or chimneys at the uppermost points of the mold cavity to allow air and gases to escape as the metal fills the mold.

Runners. Another term used for the smaller, side-branching sprues that connect the main sprue with the wax pattern.

Shock-heat. To insert a mold quickly into a hot kiln preheated to at least 1000°F, required for burning out ceramic shell molds.

Slurry. The thin, soupy mixture of a fine, insoluble material. In this case, fused silica powder suspended in colloidal silica forms the slurry into which wax patterns are dipped to build ceramic shell molds.

Slush casting. Pouring a liquified material such as melted wax or clay slip into a mold, waiting only briefly for a thin coating to be deposited on the inner surfaces of the mold, then emptying the remaining liquid.

Sprue, sprueing system. A wax rod or network of rods attached to the wax pattern. After investing and dewaxing, these rods will leave hollow channels which allow the metal to enter the mold cavity.

Stucco. The fine and coarse fused silica sands which are sprinkled onto the slurry-coated wax patterns to build ceramic shell molds; also, the act of applying the sand to the damp mold.

Vents. An alternate term for risers, which are defined above.

Viscosimeter. A device for measuring the viscosity of a liquid such as the slurry.

SELECTED BIBLIOGRAPHY

"Ancient Near Eastern Art," *The Metropolitan Museum of Art Bulletin,* XLI, 4 (Spring 1984).

Bidwell, H. T. *Investment Casting.* London: The Machinery Publishing Co., Ltd., 1969.

Cellini, Benvenuto. *The Autobiography of Benvenuto Cellini.* Translated by J. Addington Symonds. New York: P. F. Collier & Son, 1910.

Clarke, Carl Dame. *Metal Casting of Sculpture.* Butler, Maryland: The Standard Arts Press, 1948.

Coleman, Ronald L. *Sculpture: A Basic Handbook for Students.* Dubuque, Iowa: Wm. C. Brown Company, 1980.

Doeringer, Suzannah, David Gordon Mitten, and Arthur Steinberg (eds.). *Art and Technology: A Symposium on Classical Bronzes.* Cambridge, Massachusetts: The M.I.T. Press, 1970.

"Egyptian Art," *The Metropolitan Museum of Art Bulletin,* XLI, 3 (Winter 1983/84).

Eliscu, Frank. *Direct Wax Sculpture.* Radnor, Pennsylvania: Chilton, 1969.

Hauser, Christian. *Art Foundry.* Craft and Art Series. New York and London: Van Nostrand Reinhold Co., 1972.

Hitchcock, Howard. "Sculpture in the College and University Art Program." Unpublished Doctor's dissertation, Teachers College, Columbia University, 1963.

Investment Casting Handbook. Chicago: Investment Casting Institute, 1968.

Investment Casting With Nalcoag® Colloidal Silica and Nalcast® Fused Silica. Chicago: Nalco Chemical Company, n.d.

Irving, Donald J. *Sculpture: Material and Process.* New York: Van Nostrand Reinhold Co., 1970.

Jackson, Harry. *Lost Wax Bronze Casting: A Photographic Essay on This Antique and Venerable Art.* New York: Van Nostrand Reinhold Company, 1979. (Reprint of the 1972 ed. published by Northland Press, Flagstaff, Arizona.)

Mehta, Rustam Jehangir. *Masterpieces of Indian Bronzes and Metal Sculpture.* Bombay: D. B. Taraporevala Sons & Co. Private Ltd., 1971.

The Metropolitan Museum of Art. *The Great Bronze Age of China: An Exhibition from the People's Republic of China.* ed. Wen Fong. New York: Alfred A. Knopf, 1980.

Miller, Richard McDermott. *Figure Sculpture in Wax and Plaster.* New York: Watson-Guptill Publications, 1971.

*Mills, John W. and Michael Gillespie. *Studio Bronze Casting: Lost Wax Method.* New York: Frederick A. Praeger, 1969.

Proceedings of the Third National Sculpture Casting Conference, March 26, 27, 28, 1964. Lawrence, Kansas: Department of Design, University of Kansas, n.d.

Proceedings of the Fifth National Sculpture Conference, May 9, 10, 11, 1968. Lawrence, Kansas: National Sculpture Center, University of Kansas, 1970.

Ransom & Randolph Investments, Binders, & Refractories for Precision Casting. Toledo, Ohio: The Ransom & Randolph Co., 1977.

Rich, Jack C. *The Materials and Methods of Sculpture.* New York: Oxford University Press, 1947.

Savage, George. *A Concise History of Bronzes.* Praeger World of Art Series. New York: Frederick A. Praeger, 1969.

Tefft, Elden C. *Lost Wax Sculpture Foundry: Design.* Lawrence, Kansas: International Sculpture Center, University of Kansas, 1971.

———. *Lost Wax Sculpture Foundry Equipment: Recommendations and Sources.* Lawrence, Kansas: International Sculpture Center, University of Kansas, 1977.

———. *Lost Wax Sculpture Foundry Equipment & Design: Ceramic Shell Supplement.* Lawrence, Kansas: Sculpture Research Center, University of Kansas, 1983.

*Verhelst, Wilbert. *Sculpture: Tools, Materials, and Techniques.* Englewood Cliffs, New Jersey: Prentice-Hall, 1973.

*Young, Ronald D. and Robert A. Fennell. *Methods for Modern Sculptors.* San Rafael, California: Sculpt-Nouveau, 1980.

**Contains instructions for building your own melting furnace.*

LIST OF SUPPLIERS

A. D. Alpine, Inc.
3051 Fujita Street
Torrance, CA 90505
(213) 530-1541
Kilns, elements, mortar, design

A-R Products
8024 Westman Avenue
Whittier, CA 90603
(213) 695-3008
Flexible molding compounds

California Crafts Supply
1096 North Main Street
Orange, CA 92667
(714) 633-8891
Wax pens, carving waxes

Capital Foundry Equipment Co.
3227 Delta Street
Rosemead, CA 91770
Slurry mixers

The Carborundum Company
Refractories & Insulation Division
Niagara Falls, NY 14302

5900 So. Eastern Avenue
Los Angeles, CA 90040
(213) 723-4866
Fiberfrax® Ceramic Fiber Hot Board

Dick Ells Company
908 Venice Boulevard
Los Angeles, CA 90015
(213) 747-5129
Wax sprues, jewelry supplies

H. Kramer & Co.
No. 1 Chapman Way
El Segundo, CA 90245
(213) 678-4281
Everdur silicon bronze

Independent Foundry Supply
6463 E. Canning Street
Los Angeles, CA 90040
(213) 723-3266
Foundry equipment and supplies

J. F. McGaughin Co.
2628 No. River Avenue
Rosemead, CA 91770
(213) 573-3000
Microcrystalline and paraffine waxes, casting materials and supplies

Nalco Chemical Company
9165 S. Harbor Avenue
Chicago, IL 60617
Mold refractories and binders

Orange County Chemical Co.
1230 E. St. Gertrude Place
Santa Ana, CA 92707
(714) 546-9901
Chemicals for patinas

The Ransom & Randolph Co.
324 Chestnut Street
Toledo, OH 43601

2337 Yates Avenue
Los Angeles, CA 90022
(213) 728-2671
Mold refractories and binders

Thorpe Insulation Co.
24100 Broad Street
Carson, CA 90745
(213) 834-7294
Distributors of
Refrasil® fiberglass tape
Manufactured by HITCO
Materials Division
1600 W. 135th Street
Gardena, CA 90249
(213) 321-8080

INDEX

References to terms defined in the glossary are printed in boldface type. Location of illustrative material is indicated in parentheses.